Introducing Spoken Dialogue Systems into Intelligent Environments

Tobias Heinroth • Wolfgang Minker

Introducing Spoken Dialogue Systems into Intelligent Environments

Tobias Heinroth
Institute of Communications Engineering
University of Ulm
Albert-Einstein-Allee 43
Ulm, Germany

Wolfgang Minker
Institute of Communications Engineering
University of Ulm
Albert-Einstein-Allee 43
Ulm, Germany

ISBN 978-1-4899-9320-5 ISBN 978-1-4614-5383-3 (eBook)
DOI 10.1007/978-1-4614-5383-3
Springer New York Heidelberg Dordrecht London

Printed on acid-free paper

Springer is part of Springer Science+Business Media (www.springer.com)

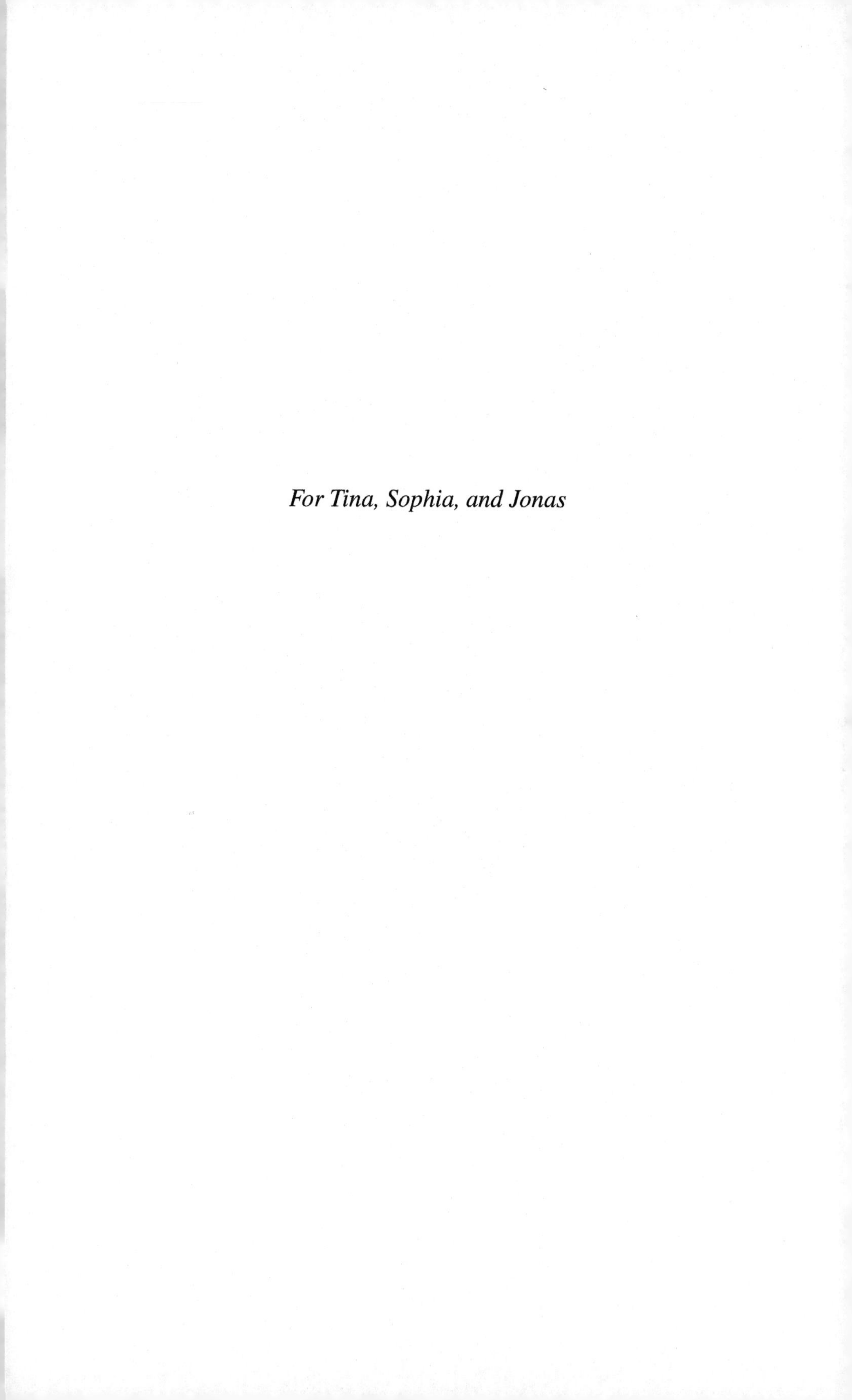

For Tina, Sophia, and Jonas

Preface

One of the main reasons for the complexity of spoken dialogue systems (SDSs) development constitutes the multi-domain and thus the multi-topic nature of real-life processes. If the application domain is not clearly defined collecting a corpus or establishing valid rules to control the dialogue flow of the SDS becomes a complex task. Within the framework of the EU-funded project ATRACO we have developed a model-based spoken dialogue manager called OwlSpeak.[1] It provides a spoken interface to an existing Intelligent Environment (IE) in real-life situations. The most important feature of the dialogue manager is its ability to pause, resume, and switch between multiple interactive tasks, hence enabling multitasking.

Our novel model-based approach allows for persistently storing the states and the structures of various spoken dialogues. Based on the multitasking capability we have defined topic switching strategies. These allow to navigate between different dialogue topics during an ongoing user–system conversation. Furthermore, we have integrated repair strategies in order to keep the dialogue coherent. We have defined mechanisms of adaptive understanding to enhance the recognition performance. Our framework also supports speaker aware dialogues and voice-based dialogue control. This enables adaptive system behaviour. Finally, we have elaborated a formal definition of dialogue descriptions that facilitate dialogue management within the dynamic IE domain. The implemented prototype is compliant with the VoiceXML dialogue description and the OWL ontology definition standards. The latter is used for knowledge representation.

During an initial system evaluation we have investigated the effects of multi-tasking on users within the spoken dialogue context. The results are twofold: users engaged in multitasking dialogues are more inclined to interact with the SDS. In turn, users who sequentially received one task after another are able to remember more facts than those who used the multitasking approach. These results led to an evaluation series focussing on different assistive dialogue strategies. They may be applied to switch the dialogue focus to a different topic (and afterwards back

[1] http://sourceforge.net/projects/owlspeak/

to the original one). The underlying idea is to guide the user by alerting him of possible dialogue interruptions. After a sub-dialogue is being processed, the original dialogue is re-introduced by reminding the user of the main topic. The analysis results indicated that the sophisticated explanation strategy performs best. Notably, the applied dialogue strategy also had a measurable and significant influence on the overall dialogue quality.

A social evaluation has been conducted within an existing IE revealed qualitative results and a positive learning process the subjects went through during the three successive evaluation sessions. However, the prototype generated dialogues that were too rigid and not sufficiently intuitive. Considering the high motivation of the subjects and the eagerness with which they controlled the IE via speech, we have investigated ways to enhance the understanding capabilities of OwlSpeak. The main goal was to render the interface more intuitive. Hence, we have evaluated different mechanisms to solve this issue whilst keeping the complexity of the domain models low. We have discovered that, especially for command-and-control dialogues, semantic strategies that enhance the understanding capabilities are the most promising approaches. A further evaluation covered the issue of how to cope with errors occurring during spoken human–machine interaction. Therefore, we have compared three strategies ranging from a simple re-prompt to a more complex self-repair strategy. The main outcome of this evaluation is the strong dependency between the choice of an appropriate repair strategy and the user characteristics. The subjective rating of experts differed significantly from the rating of novices. This underpins the importance of user-related information for dialogue management.

The theoretical foundations of a working ontology-based spoken dialogue description framework, the prototype implementation of the ASDM, and the evaluation activities that have been conducted as part of this work contribute to the ongoing research on spoken dialogue management by establishing the framework of model-based Adaptive Spoken Dialogue Management.

The research leading to our results has received funding from the European Community's 7th Framework Programme (FP7/2007–2013) under grant agreement n°216837 and from the Transregional Collaborative Research Centre SFB/TRR 62 "Companion-Technology for Cognitive Technical Systems" funded by the German Research Foundation (DFG).

Ulm, Germany

Tobias Heinroth
Wolfgang Minker

Contents

List of Figures

List of Tables

Acronyms

AIE	Ambient Intelligent Environment
API	Application Programming Interface
AS	Activity Sphere
ASDM	Adaptive Spoken Dialogue Manager
ATRACO	Adaptive and TRusted Ambient eCOlogies
DTMF	Dual-tone multi-frequency
EMMA	Extensible MultiModal Annotation markup language
FTA	Fuzzy Task Agent
GUI	Graphical User Interface
HCI	Human–Computer-Interaction
HTTP	Hypertext Transfer Protocol
IA	Interaction Agent
IE	Intelligent Environment
MVC	Model-View-Controller
MVP	Model-View-Presenter
NLU	Natural Language Understanding
OWL	Web Ontology Language
POMDP	Partially Observable Markov Decision Processes
PTT	Push-To-Talk
SM	Sphere Manager
SDM	Spoken Dialogue Manager/Spoken Dialogue Management
SDO	Spoken Dialogue Ontology
SDS	Spoken Dialogue System
UML	Unified Modelling Language
UPnP	Universal Plug and Play
UsiXML	USer Interface eXtensible Markup Language
VoiceXML	Voice Extensible Markup Language
VUI	Voice User Interface
XIML	Extensible Interface Markup Language
XML	Extensible Markup Language

Chapter 1
Introduction

During the past few decades, the development of Spoken Dialogue Systems (SDSs) has advanced significantly due to increasing miniaturisation of electronics combined with reduced costs. This paved the way for the development of specialised speech recognition and synthesis algorithms. The emergence of powerful mobile devices along with the increasing accessibility of the Internet has also enabled the development of a multitude of speech-related applications. In the cellular telephony arena SDSs are already quite sophisticated. Apart from information retrieval, call routing, and transactional applications, new technical support systems for customers have become more widely available. Such automated agents help callers to, for example, solve Internet-related problems or resolve technical issues with various devices. In automotive applications such as route guidance or control of entertainment systems, a plethora of spoken command systems are available in which—more or less regular—spontaneous speech is accurately understood. Two key technologies have facilitated these advancements: voice recognition (Automatic Speech Recognition—ASR) and speech synthesis (Text-to-Speech—TTS). Apart from these technologies, an SDS also performs linguistic and semantic analysis, text generation, and contains a Spoken Dialogue Manager (SDM) that constitutes the behaviour and the conversional characteristics of the system.

Depending on the desired method of speech recognition (i.e. free text, keyword-based or dual-tone multi-frequency signalling (DTMF)) as well as on the commands the system should be able to handle, there are several differing requirements. When using a mobile phone for example, complex modes of communication such as *negotiating* or *discussing* are usually not necessary: a phone should merely understand commands such as "Call Peter in the office". However, even such commands are not as straightforward as they may appear at first glance. For example, the phone should not immediately call Peter if "...you might call Peter in the office and ask him..." is uttered. Commercially available products such as the personal assistant SIRI [developed as a byproduct of the DARPA-funded project CALO (Duong et al. 2005; Gil and Ratnakar 2008; Gervasio and Murdock 2009)] utilise Push-To-Talk (PTT) mechanisms to detect valid user utterances.

T. Heinroth and W. Minker, *Introducing Spoken Dialogue Systems into Intelligent Environments*, DOI 10.1007/978-1-4614-5383-3_1,

PTT, however, is not optimal for use in personal assistant systems. In practice, users would probably prefer a system that is always "listening", able to recognise system-directed input without requiring a button press. Other factors such as personalised voice detection and multi-party conversations are additionally important considerations for this endeavour. Once the user's speech has been detected, large-scale semantic analysis (usually performed on the server side) can be performed in order to interpret the input. Despite recent progress in this area, there are still many unresolved issues regarding inherent dialogue management and conversation control. In particular, the formal description of dialogues and the complexity of discourse are scientific topics of interest. In describing the current state of the art of SDSs, the following quote by Hemingway highlights one of the major remaining research challenges:

It takes two years to learn to talk, but fifty to master silence.

Even though we now have SDSs for desktop computers, telephony agents, mobile phones, and infotainment systems, it will still take some time before computational systems will be able to discern the appropriate time to not respond (keep silent). Aside from the question of *how* to respond, there is the question of *what* words to speak to a machine.

In this document we focus on the management of spoken dialogues within Intelligent Environments (IEs). Cook et al. (2006) define an IE "[...]as one that is able to acquire and apply knowledge about its inhabitants and their surroundings in order to adapt to the inhabitants and meet the goals of comfort and efficiency". Within the ATRACO project[1] we have defined IEs using the term *ambient ecologies* (Goumopoulos and Kameas 2009). Here we conceptualise a space in a similar way but focus on the support of everyday user activities. These definitions imply that the classical *single-task* approach to SDS is inadequate. An IE or an ambient ecology generally cannot be grasped as a single application providing a predefined set of functionalities. Instead the spoken interface must be sufficiently flexible for use by a variety of devices, services, knowledge sources, and user inputs. Notably, all entities involved vary continuously depending on the status of the IE and the current task(s)—even during an ongoing user–system conversation.

Hence, an SDM for IEs must be able to simultaneously handle *multiple tasks* covering different aspects of communication. These aspects may be categorised as (1) **command-and-control** dialogues, (2) interaction for information **retrieval** (on the part of the user), and (3) questions regarding information **gathering**. The SDM must be capable of addressing a meaningful combination of these. Here we propose an approach referred to as *Adaptive Spoken Dialogue Management* (ASDM). This technology addresses the need for enhanced dialogue control in fulfilment of specific IE requirements for a natural speech interface. In the following three

[1] This work has been carried out in the ATRACO project within the European Community's Seventh Framework Programme (FP7/2007–2013) under grant agreement n°216837.

paragraphs we discuss examples of all three aspects to illustrate the requirements and to highlight the use of adaptation within the IE context.

If, for example, a user must control 12 lights within an IE, a graphical user interface would most likely consist of 12 buttons. A graphical layout with less than 12 configurable buttons would also be possible but probably not as user-friendly. An SDS would also need at least 12 commands for controlling the lights but the user would not have to perceive all of them. Instead he could just ask the system to "[...] switch on light A" and would not have to cope with a graphical control panel. Furthermore, in case ambiguities occur, the SDS could check back on which light the user wants to switch on. Within the scientific area of IEs, SDS technologies thus offer one of the most efficient and natural interfaces between humans and computer-based systems.

Since IEs consist of networks of various different components such as sensors, actuators, and processors, they automatically exchange information about themselves and their surroundings without human intervention. Thus a user can provide input to component A that analyses the request and provides the new information to other (related) components that may in turn execute this request, or pass the information to other components as well. Within the framework of IEs for many tasks and especially for the command-and-control of devices or services mentioned above, proactive behaviour (warnings, information, etc.) and the intuitive input on the part of the user make speech a promising modality. In particular for the elderly, disabled, and/or people with serious health issues an SDS would be very useful since it provides a centralised and highly accessible natural language interface. This interface however would only be usable if the dialogue and the state of the IE correlate—that is, if the SDM *adapts* to the state of the IE.

A further very complex issue is related to the retrieval of information via spoken dialogue on part of the user. The following example illustrates the major difficulties. The user is out shopping and calls the IE at home to ask "Do we have enough milk?". This example reveals an intrinsic problem of speech: its inexactness. If the user has enough milk the correct system answer would be "Yes"—even if, for example, the milk has soured. We can address the issue of inexactness by asking a more precise question: "Do we have enough edible milk?" Of course, it would be more convenient if the IE would independently discover whether or not the milk is edible. In this case the correct system response would be: "Yes, but it has soured". This last example reveals a further challenge regarding the realisation of real-life SDSs within the framework of IEs: its complexity. We argue that one possibility for reducing the complexity is *adaptation*: the dialogue decision logic has to be able to select the correct information that is specifically needed in order to render a dialogue.

Not only does the user need to retrieve information from the system. The system also needs to attempt to gather information from the user. The system should be optimally designed to ask the user questions in a precise and concrete manner. Here, the issue of speech complexity still remains a major challenge. A related difficulty is determining the correct time for the system to ask a question of the user. The urgency of a specific item of information must be taken into account here. Considering single-user scenarios, the user may already interact with the system via voice. In this document we discuss the major issues that arise if the system interrupts an ongoing dialogue. A voice interface that is used in multi-party scenarios also has to take the ongoing inter-user conversation into account. Again, we have to highlight the necessity of an *adaptive* SDM that is able to integrate the contextual information into the decision logic in order to provide a spoken interface to an IE.

How much information needs to be taken into account so as to allow for a meaningful dialogue between the user and the system? How could this information be modelled to be both readable by computers and easily assimilated into an ongoing dialogue? An SDS and most notably an SDM must be capable of accomplishing these activities, depending on the scenario and the specific task to be achieved. It must be capable of negotiating the balancing act of understanding and interpreting while reducing the complexity as much as possible. The examples above demonstrate that the development of a more intelligent and therefore adaptive SDS requires more than advanced voice recognition and speech synthesis. To a greater degree an ASDM that serves as the core of an SDS and controls dialogue discourse and integrates the knowledge of the IE is sorely needed.

1.1 Problem Setting

The above examples illustrate several scenarios that a well-designed SDS and more specifically an ASDM must be able to handle within the IE framework. In this section we describe the three major types of spoken dialogues we address in this document: command-and-control, information retrieval (on the part of the user) and information gathering (on the part of the system). The W3C standardized VoiceXML description language is widely used for this today (Oshry et al. 2007). The idea behind this approach is to simplify the development of dialogues by providing a model description of the conversation to be expressed. Thus a specific description may be used to define the structure of a particular dialogue in a robust manner.

VoiceXML is limited to system-initiative and mixed-initiative dialogue layouts. System-initiative dialogues are typically controlled by the system. In practice this means that the user is only able to provide input to a previously stated question, which is one portion of the SDS. As illustrated in the previous section this functionality realises dialogues for information gathering. Mixed-initiative dialogue layouts can be seen as an extension of this rigid system behaviour. Utilising such a layout allows the user to provide more information than the actual question asks for. This most important feature of mixed-initiative dialogues is called over-answering. However, even this enhancement primarily only provides a framework for information gathering on the part of the system.

Both command-and-control and information retrieval dialogues require that the dialogue waits until the user *wishes to provide spoken input*. VoiceXML however lacks such functionality as unlimited loops, task-specific pauses, and dialogue resuming. On one hand, VoiceXML provides ease-of-use for dialogue developers.

On the other hand, its expressiveness remains limited regarding more complex structures such as task-oriented dialogue flows. Furthermore, VoiceXML is not able to persistently store and therefore describe a specific state of a dialogue. Thus dialogue strategies such as pausing and resuming of parallel tasks can hardly be implemented. Keeping these limitations and the special IE requirements we face in mind, we summarise the major issues as follows:

Command-and-control dialogues. In principle it is straightforward to implement dedicated spoken dialogues for specific devices or services (i.e. entities). Within the IE framework, however, the available entities continually change. Thus it is intractable to define dialogues (e.g. using VoiceXML) for controlling groups of entities (heating, doorbell, and lights): If one of these entities becomes no longer available, a new dialogue must be defined. A more efficient approach would be to consider the varying entities as multiple tasks that might be activated or deactivated during a session (i.e. while the user interacts with the IE). Therefore, we emphasise the presently unsolved issue of *multitasking* in spoken dialogues and present a possible solution for this in the following.

Information retrieval dialogues. Similar to command-and-control dialogues, the SDS must wait until the user utters a specific request. Here the user does not state a command, but rather asks for (task-related or even unrelated) information. As for command-and-control input on the part of the user, the information retrieval may also be requested during an ongoing dialogue. A typical example for such a dialogue is a counter question that may first be required before the user may proceed with the original dialogue. Furthermore the user has to be able to ask for help in case he is not sure how to proceed with the current dialogue. Here we emphasise the necessity to *pause* and *resume* dialogues and to *switch the focus* from one to another. Within this context a further challenge is to persistently store the dialogue state. In the following section we will point out our proposal for a possible solution to this.

Information gathering dialogues. In contrast to the scenarios described above, information gathering dialogues allow the system to decide when to ask a specific question, and therefore to prompt the user for a response. This kind of dialogue closely relates to the original purpose of VoiceXML. However, the focus of this document is on the interaction within IEs and not on telephony-based dialogues. Hence, an important issue is *when* to initiate such an information gathering dialogue. In the present work we point out the necessity of *prioritising* dialogues. Furthermore we assume that the IE is able to *activate* dialogues depending on contextual information. This last point is an important prerequisite to a successful integration of the ASDM into the IE.

Multitasking, persistent *dialogue storage*, dialogue *prioritisation*, and the *integratability* of the SDM into the IE are in our view fundamental requirements that must be fulfilled in developing an SDS that behaves adaptively and eventually learns *when to be silent*. We outline and evaluate mechanisms and methodologies to meet the above-mentioned requirements. In the following section we briefly outline our study procedures and objections.

1.2 Proposed Solution: Adaptive Spoken Dialogue Management

Our solution to the problem setting is a modular framework that we call ASDM. Our proposed solution addresses three aspects of ASDM as follows:

Theoretical. We describe the classification of spoken dialogue adaptation. Three stakeholders influence the spoken dialogue: the user, the SDS, and the IE. Each provides specific levels of adaptation that an ASDM must be able to handle. We define Behavioural Adaptation and Emotional Adaptation as two levels that directly relate to the user. Dialogue Strategy Adaptation and Speech Adaptation refer to the capabilities of the SDS. Device Adaptation, Event Adaptation, and Task Adaptation describe the behaviour of the ASDM with respect to the environment. The proposed classification is a valuable outcome of our work as it will provide a thorough catalogue of the requirements the ASDM must fulfil given the changing IE domain. As a direct consequence we have decided to introduce a model-based approach to spoken dialogue management that leads to a domain independent and generic architecture.
A further important theoretical contribution is that we define a formal framework for persistent spoken dialogue description. To the best of our knowledge there is no approach available that combines both the dialogue state and its structure. Either structural descriptions [e.g. VoiceXML (Oshry et al. 2007)] or dialogue state-related approaches [e.g. The Information State (Traum and Larsson 2003)] have been investigated. However, the combination of the state and the structure that we propose allows for reliable dialogue storage thereby enabling the ASDM to perform multitasking dialogues. We present the formal definition of a Spoken Dialogue Ontology and describe how the ASDM utilises this kind of dialogue data.

Practical. In the proposed approach to develop an ADSM, we combine an expressive description technology, namely OWL ontologies (McGuinness and van Harmelen 2004), with a standardized VoiceXML input and output layer. This layer is maintained by a dialogue logic that turn-wise generates spoken dialogue snippets. Depending on the status of the underlying OWL knowledgebase, these snippets represent a currently valid dialogue. The status of the underlying dialogue model itself depends on contextual information, i.e. on the status of the IE. The idea is to develop a model that is sufficiently generic to describe dialogue domains of various flavours such as command-and-control structures for device control or more complex user-initiated dialogues for information retrieval and gathering. We assume that the gap between these different types of dialogues can only be filled if the system is adaptive. Our prototype system, called OwlSpeak, provides adaptive spoken dialogues by utilising OWL models and broadens or shortens these knowledgebases depending on the interactive task(s) to be carried out. The layered architecture depicted in Fig. 1.1 facilitates the provision of multitasking because the ASDM is able to directly react to changes that have been applied to the model—thus the system behaves adaptively.

Fig. 1.1 The underlying idea of a layered ASDM architecture

By utilising specific strategies for switching between different interactive tasks, the conversation between user and system may consist of main dialogues interspersed with sub-dialogues with the option to use commands to trigger specific entities such as display lights. OWL ontologies do not only encode the dialogue description the same way as, for example, VoiceXML does, but also persistently encodes the current status, i.e. the progress of the dialogue. Since the dialogue model strictly distinguishes between different dialogue components, a rigorous prioritisation of the model can be achieved. The open architecture and the clear definition of the dialogue model allows for a flexible and straightforward integration into an existing IE (e.g. the ATRACO system). Furthermore, the prototype is available as open source for the broader scientific community.
Several approaches of enhancing the ASDM have also been investigated. On one hand, we have designed specific dialogues using strategies such as multitasking support and repair methods as part of the experiments and evaluations described in the following paragraph. On the other hand, we have implemented techniques allowing the use of keywords in the event that the user's speech has not been correctly understood before. The integration of a semantic-lexical knowledgebase has also significantly enhanced the capabilities of the prototype.

Experimental. The general functioning of the ASDM has been tested during an initial system evaluation. We also investigated the effects of multitasking on users within the spoken dialogue context. The results are twofold: users who were engaged within multitasking dialogues were more inclined to interact with the SDS. These users received several reminders during an ongoing dialogue. In turn, a second group of users received the reminder one after another (serially) after the main dialogue has been completed. These users were able to retain more facts than those who used the multitasking approach.
This unexpected result that revealed benefits for both approaches in turn led to a second evaluation series. During this series we focussed on different strategies that can be applied in order to assist the user if the ASDM must switch the dialogue focus to a different topic (and afterwards back to the original one). The analysis results indicated that a sophisticated explanation strategy outperformed other strategies by alerting the user that the dialogue has to be interrupted. Having processed the sub-dialogue, the original dialogue is reintroduced by reminding the user of the topic (e.g. dinner preparation). Notably, the applied dialogue strategy also had a measurable and significant influence on the overall dialogue quality. The explanation provided by the system led to a higher performance of the entire dialogue.

A social evaluation that has been conducted within an existing IE in realistic conditions revealed qualitative results that influenced our research. The subjective results relate to a positive learning process that the subjects went through during three evaluation sessions. However, the prototype that has been integrated seemed to generate dialogues that were too rigid and not sufficiently intuitive. Considering the high level of motivation and the eagerness with which the subjects control the IE via their speech, we investigated ways to enhance the understanding capabilities of the ASDM.

The main aim was to render the interface more intuitive whilst keeping the complexity of the dialogue description the same. Hence, during a further user evaluation we explored different mechanisms that could be integrated such as the use of keywords. We also applied the Levenshtein distance to blur the detected user input and integrated a semantic-lexical knowledgebase. We discovered that semantic strategies that enhance understanding, especially for command-and-control dialogues, are the most promising approach. The main reason for this was the intuitive naming of the devices, which exceeded the initially applied system commands.

An additional evaluation session addressed the question of how to cope with mistakes that occur during a spoken human–machine interaction. Therefore we compared three strategies ranging from a simple re-prompt to a more complex self-repair strategy that tried to guess the correct user input. The main outcome of this evaluation was that the particular repair strategy to be applied strongly depended on user characteristics. The subjective rating by experts differed significantly from the rating performed by novices. This underpins the importance of user-related information that needs to be taken into account by an ASDM.

Finally we have proven the practical use of the OwlSpeak ASDM by conducting a scalability analysis. We tested the performance of the system on several measures (e.g. start-up time, processing time). We then tested how these measures are influenced when the number of devices and services grows from 1 to 100. The most important outcome was that the system itself is scalable. However, the automatic ambiguity detection that cross-compares the user's input with all commands that the devices accept is a bottleneck. The current implementation of the ambiguity detection performs within acceptable timing for up to 30 devices.

In the following section we provide the outline of the remainder of this document.

1.3 Document Structure

This document consists of six chapters. The present chapter provides the introduction and the motivation behind our study. The second chapter reports on related work that has been conducted within the area of Spoken Dialogue Management and Intelligent Environments. The third chapter focuses on novel approaches to ASDM that have been developed as a part of our work. The fourth chapter reports on the technical implementation of the dialogue management framework. The experiments and evaluation sessions that have been conducted are discussed in the fifth chapter. Chapter 6 concludes the document and discusses future research directions.

In the next chapter we provide the related work regarding SDSs and the fundamentals of IEs. We also focus on the interaction within IEs before we present general approaches to SDM. Enhanced SDM methodologies are described in Sect. 2.5. Chapter 3 provides a definition of ASDM within IEs. Section 3.5 illustrates the most important requirements on the basis of an application scenario. The most important characteristics of the functionality of the ASDM are described in Sect. 3.6. Following that, the system fundamentals of the proposed prototype—the OwlSpeak ASDM—are explained in Chap. 4. Here, we focus on the proposed model, the theoretical background, and the logical structure of the ontology that is used to define spoken dialogues. Details about the architecture and the implementation of the ASDM are also explained.

Chapter 5 begins with a brief description of the various experiments that have been carried out. Section 5.1 reports on the first user evaluation that proved that the system can be utilised as a realistic test bed for further investigations. Section 5.3 presents the results of a comparison of different topic switching strategies that can be applied to solve several issues of spoken dialogue multitasking. The qualitative results of the social evaluation that has been conducted as part of the EU-funded project ATRACO are analysed in Sect. 5.5. Our work on advanced understanding mechanisms has also been evaluated: the outcome is presented in Sect. 5.6. Section 5.4 presents the results of the comparison of different repair strategies and the impact they have on a spoken dialogue. Finally, the results of a scalability experiment are presented in Sect. 5.2. The third part concludes with a summary of the evaluation activities in Sect. 5.7. In Chap. 6 we discuss the scientific contribution and present conclusions. Section 6.3 describes some proposed future directions the research might take and presents an outlook on future work. The appendix which follows consists of a description of the functionality of the command-and-control dialogues that have been implemented within the ATRACO prototype (Appendix A), the OWL definition of the Spoken Dialogue Ontology (Appendix B), and UML diagrams describing the most important algorithms (Appendix D). Finally Appendix E presents an exemplary questionnaire used in the evaluation sessions.

Chapter 2
Background

In this chapter we discuss the background and the related work of ASDM within the context of IEs. In Sect. 2.1 we explain the general functioning of an SDS. The underlying idea of an IE is explained in Sect. 2.2. The IE approaches realised within the ATRACO Project serve as examples. In Sect. 2.3 we describe prior work in the field of (spoken and multimodal) interaction within IEs. Section 2.4 focuses on a specific part of an SDS: the Spoken Dialogue Manager (SDM). Several approaches toward developing this component have been implemented in the past. We give an overview on all directions in general and illustrate each with an example. Section 2.4 is divided into three parts: the first part is dedicated to state-machine-based approaches, the second part to stochastic methodologies, and the third part to plan- and Information State-based systems. In Sect. 2.5 we present several approaches to enhancing the performance of the SDM. Furthermore, we discuss how these approaches influenced our work. By introducing our own approach we conclude this chapter in Sect. 2.6.

2.1 Spoken Dialogue Systems

A Spoken Dialogue System (SDS) is a computer-based system that enables a user to bilaterally communicate via spoken language with a machine (hardware and/or software). Figure 2.1 shows an architectural overview on a standard SDS. The three most important layers are the acoustic front-end, the semantic layer, and the logical layer. Speech recognition and speech synthesis modules constitute the acoustic front-end. This layer is usually accessed by the user via microphone(s) and speaker(s). The speech recognition process consists of an analysis that extracts a set of features from a discrete speech signal. These features are then correlated with graphemes or words provided as part of a language model. The synthesis layer provides the reverse direction: appropriate mappings are used to transform graphemes or words into acoustic signals. In order to realise the bidirectional connection between the acoustic front-end and the logical layer an SDS provides

T. Heinroth and W. Minker, *Introducing Spoken Dialogue Systems into Intelligent Environments*, DOI 10.1007/978-1-4614-5383-3_2,

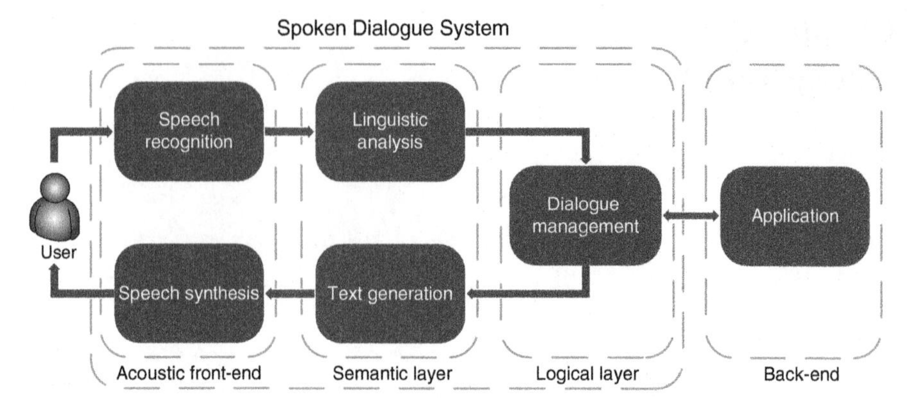

Fig. 2.1 Architectural overview on a standard spoken dialogue system

a semantic layer. It consists of two modules: the linguistic analyser and the text generator. The main concern of the linguistic analyser is to provide a mapping between the output of the speech recogniser and the semantic values described as part of the logical layer. The semantic values must be computationally readable and valid. Reversely, the text generator must provide human understandable text. Out of the machine readable semantic values this text can be rendered to audio signals. The third layer, the logical layer, is constituted by the SDM. In conventional SDSs the main role of the SDM is to link the semantic values provided by the linguistic analyser with the semantic values that should be interpreted by the text generator. Hence, the SDM provides the connection to the application, which is the user's counterpart during the spoken conversation. This application constitutes the so-called back-end.

In practice it can be individually constructed using different modes and modalities for different purposes (e.g., information retrieval, task execution, user support). In Chap. 4 we present an alternative definition of SDM. We propose to broaden the focus of the logical layer by adding several aspects of the linguistic analysis, the text generation, and the back-end to the original role of the SDM. As a result we compose the ASDM prototype system OwlSpeak. In the remainder of this document we use the terms *OwlSpeak* and *Adaptive Spoken Dialogue Manager* as well as the abbreviation *ASDM* interchangeably. In summary, an SDM can be seen as the central component of an SDS. Therefore, it seems natural to design the complete SDS as a framework based on the SDM. In the following we provide insights into such a framework: the Olympus SDS that has been established by the Carnegie Mellon University (Bohus et al. 2007). The Olympus architecture defines a set of components that may be utilised to implement an SDS:

- The ASR *Sphinx* is used to recognise the input uttered by the user (Huerta 2000). Sphinx is a statistical recogniser based on Hidden-Markov Models (HMMs).
- The natural language understanding is done by *Phoenix*, a robust parser based on context-free grammars (Ward and Issar 1994).

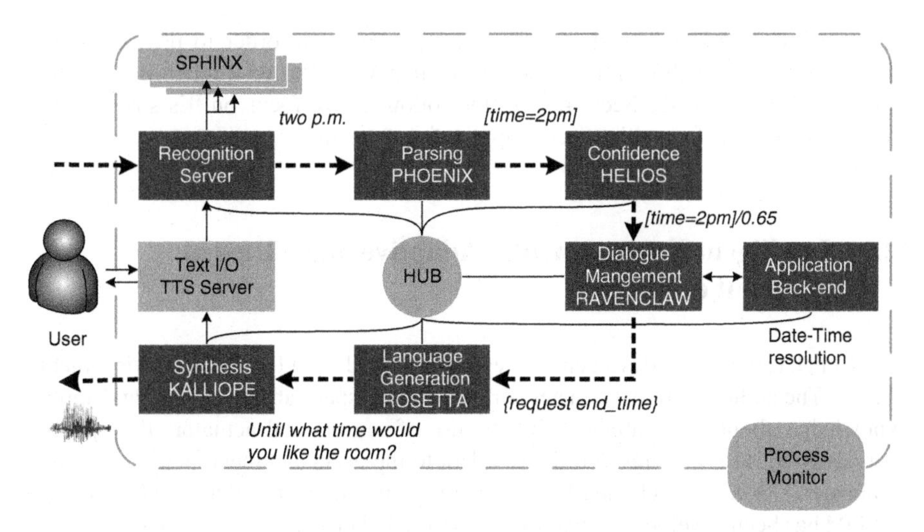

Fig. 2.2 Overview on the Olympus SDS Framework Architecture, cf. (Bohus et al. 2007)

- The architecture provides a confidence annotator that integrates multiple sources of confidence of a particular understanding: *Helios* (Bohus and Rudnicky 2002).
- The *Apollo* low-level interaction manager takes care of exact timing of start and end of utterances and the handling of interruptions (Raux and Eskenazi 2007).
- Natural language generation uses the *Rosetta* template-based generation system that has been introduced by Oh and Rudnicky (2000).
- The speech synthesiser interface proposed by the architecture is *Kalliope*. It supports several established TTS engines (SAPI 5, etc.).
- The dialogue management is handled by *RavenClaw* (Bohus and Rudnicky 2009). This dialogue manager runs fully task independent and can be seen as the core component of Olympus.
- The communication handling between the different components is provided by the MIT/MITRE *Galaxy Communicator architecture* (Seneff et al. 1998).

Figure 2.2 shows an overview on the Olympus architecture. The strict distribution of the various components offers several advantages: the system is modular since all components may be replaced by updated versions or even by different components providing similar functionality. The SDM Ravenclaw follows an agent-based approach. This means that several concurrent agents compete for fulfilling their specific dialogue goal (e.g., flight arrival time, destination) given the actual user input. A task definition language defines how an agent has to react to a specific input. By selecting the most promising agent (using a confidence measure), the SDM decides how to proceed with the conversation.

A main drawback of such a distribution is that it is inflexible. In case the system undergoes a major change, the definitions and the settings of various components must be updated. However, the ability to react to domain changes is an important requirement of *adaptive* system behaviour. Therefore, in order to provide more flexibility, we have chosen a central part of the ASDM to be domain-dependent: the dialogue model (see Sect. 4.3). In the following we focus on IEs since many requirements arise from this specific application domain.

2.2 Intelligent Environments: Adaptive and TRusted Ambient eCOlogies

Cook et al. (2006) provides a general description of what an Intelligent Environment (IE) is. The authors define it as a networked physical space able to acquire and apply knowledge about its inhabitants. By the use of sensors and actuators the system perceives the surroundings in order to adapt to the users and meet their goals. The main aim is to achieve a higher level of comfort and efficiency. Since the OwlSpeak ASDM has been developed as part of the EU-funded Project "Adaptive and TRusted Ambient eCOlogies" (ATRACO), we describe the properties of an IE on the basis of ATRACO. One of the most important features of OwlSpeak, its multitasking capability, arise from the underlying ATRACO ideas. Nevertheless, the ASDM may also be integrated into other types of IEs (e.g., Intille et al. 2005; Mozer 2005; Kientz et al. 2008). The aim of the ATRACO project is to contribute to the realisation of Activity Spheres (AS) that are established within so-called ambient ecologies. An ambient ecology consists of a set of devices located in close proximity and several corresponding services. Both devices and services may communicate and collaborate with each other, the environment and the people (Goumopoulos and Kameas 2009). The overall objective of ATRACO is to lay the foundations for the development of a new range of concepts, models, components, architectures, and guidelines that underpin the development of such ambient ecologies. The resulting conceptual framework consists of specific concepts implemented as a multi-layered ontology. This ontology hierarchically describes the basic and higher-level system behaviour and allows for a novel interaction metaphor. There are two main research aims: Research on adaptation and on heterogeneity. The former focuses on defining and understanding different types of adaptation at system level. These types arise from or depend on the behaviour, the goals, and the adaptability of individual parts of a specific ecology. This also includes the interactions amongst these parts within the environment. The latter focuses on novel methods of managing devices and services that do not provide a heterogeneous interface such as a common protocol or a middleware. Both research aims also describe the main issues the ASDM faces: it must adapt on different levels (as described in Chap. 3) and must be integrated into a framework that consists of varying entities. As defined in (Goumopoulos and Kameas 2009) the relevant properties of an ambient ecology are:

- It is distributed since the various entities (i.e., devices and services) within the IE are also distributed.
- It is composed of heterogeneous hardware and software components.

- It is dynamic in its structure and in the configuration of interactions amongst its entities.
- It is reactive to changes in its environment and to the interrelationships amongst its entities.

The distribution of heterogeneous entities, the inherent dynamic, and the reactiveness of the system requires adaptation. This is also a major requirement of our approach to SDM. Technically, an ambient ecology resides within an IE. It consists of entities (i.e., users, agents), devices and services, and ontologies used as knowledgebases. Thus, an AS is both the semantically rich description of the resources required to achieve a specific user aim and its instantiation within the context of a specific ambient ecology. Therefore, multiple ASs, each corresponding to a separate aim or goal, may be instantiated concurrently. In this case the ASs would use the resources of the same ambient ecology at the same time. Each AS is regarded as an autonomous instance of ATRACO and is supported by an independent ATRACO system. Hence, all spheres adopt the same ATRACO architecture. An AS consists of:

- A description of a goal defined as a hierarchical task model.
- Users, devices, services (i.e., the entities within the IE), each providing its own ontology. This ontology stores both the description of the entity and its current state.
- A set of policies to define privacy rules, derive modes and modalities of interaction, etc.
- A sphere ontology used as primary knowledgebase for a specific AS.
- A set of agents and other software components such as the sphere manager. Here we introduce the Interaction Agent (IA) since the ASDM has been integrated into this component.

The ATRACO sphere manager is responsible for creating, managing, and dissolving ASs. Various agents are responsible for resolving conflicts, interacting with the user and in general realising the user tasks. In general, a task model in ATRACO describes all tasks that have to be performed in order to reach a specific goal (e.g., to prepare a dinner) (Van Welie et al. 1998). The task model may be decomposed in order to create, for example, a tree. Once a specific task cannot be further decomposed into subtasks, the individual subtasks can be realised (i.e., can be carried out). The sphere ontology is managed by the ontology manager. This ontology results from merging the local ontologies of all entities that are required to achieve a specific goal. It contains all goal-related knowledge and information. The ontology manager informs the various agents, when the sphere ontology is modified. Thus, the agents can directly take advantage of a homogeneous information pool. Agents, devices, and services autonomously maintain and update their local ontologies. As a result, the sphere ontology, being the result of aligning the local ontologies, reflects the most recent state of the AS. Figure 2.3 depicts an AS that incorporates different knowledge sources. The example demonstrates the alignment of a set of devices (the TV, a window, and the lights) and a set of services providing information about the foods that are available in the kitchen.

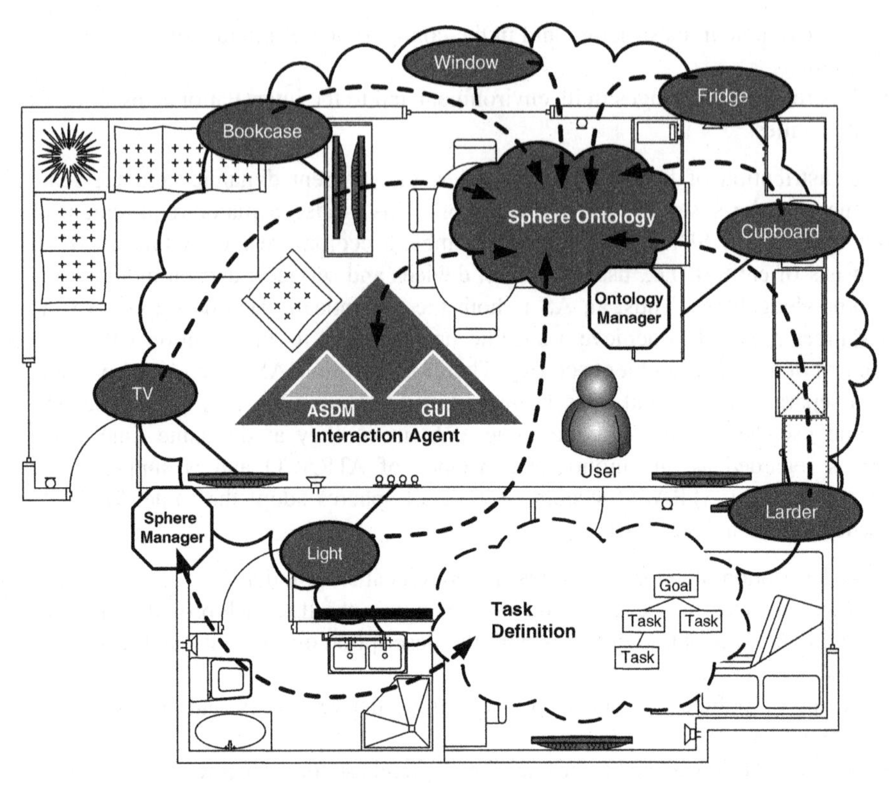

Fig. 2.3 High-level view of an ATRACO instance

The ASDM participates within an ATRACO AS as an autonomous entity that is part of the IA. The IA acts as a multi-modal interface provider. Its main objective is to select an appropriate set of interfaces. It is considered that the selected interfaces are beneficial within the current context. We call these interfaces *mediators*. The ASDM itself is such a mediator and serves as an adaptive voice interface to the entities affected by the user (i.e., to the entities that are part of the AS and that can be accessed via voice). The IA decides, depending on predefined rules, which entities currently can be addressed. The user interacts with the system in a *free-play* mode, meaning that the dialogue flows are only partly prescribed. Analogously to all ATRACO tasks, interactive tasks may also be part of the task model. These interactive tasks are decomposed into a tree as suggested by Paternò et al. (1997). The IA is able to dynamically allocate two main types of interactive tasks to different mediators. The mediators provide the interfaces by utilising various devices and services. As defined in (Pruvost et al. 2011) the two types of interactive ATRACO tasks are:

- *Control tasks*, which imply the use of interfaces for a persistent control of the IE. Such dialogues are also known as *command-and-control* dialogues.

- *Dialogue tasks*, which are used by the system to provide the user with information (i.e., information retrieval) or to ask for information (i.e., information gathering).

The ASDM adapts the spoken interface to the current set of interactive tasks that are requested for realisation by the IA. Since the interaction capabilities available within an IE are unknown before runtime, the IA and therefore the ASDM behave *adaptively* by composing the user interfaces on the fly (i.e., during runtime). A similar handling of tasks has been described by Niezen et al. (2010) as part of the SOFIA project.[1] As depicted in Fig. 2.3 the AS integrates the bookcase and the food cupboard. Based on the application scenario, the aim of this AS is to support a dinner preparation (cf. Sect. 3.5). The various entities provide their own knowledgebases, the local ontologies. As an example, the bookcase ontology provides information about the available cookbooks and cupboard ontology lists the available food. This information is aligned so as to form the sphere ontology. This unified knowledgebase *homogeneously* provides all information of the *heterogeneous* devices and services. Depending on what entities the user wishes to interact with, the IA decides which mediators to use. If the AS allows for using spoken dialogue to control specific entities, the IA would dynamically trigger the ASDM to activate the appropriate dialogue models. These dialogue models are encoded as Spoken Dialogue Ontologies (SDOs) (see Sect. 4.3). As a result, the ASDM generates a spoken dialogue taking the current state of the AS into account. As mentioned, from the IA's point of view the ASDM acts as a mediator. Hence, it provides the interfaces and adapts them to the current context. More details on the IA have been published in Bellik et al. (2010).

The IA must be aware of all information flows between the user interfaces and the ATRACO system. Hence, we decided to implement all mediators [the ASDM and a Graphical User Interface (GUI)] as subcomponents of the IA. All event flows related to human–computer interaction (HCI) are redirected to the respective mediator by the IA. Furthermore, all events emitted by the ASDM are also passed to the IA. In turn, the IA sends the events to the core component of each AS, the sphere manager. The GUI mediator provides a widget-based framework. This framework can easily be adapted by adding or removing widgets for different tasks. However, for the following reasons the realisation of spoken dialogue interface adaptation is more complex:

- Spoken language is not as precise as a GUI usually is. For example, if the user utters "lights on" it is not defined which light he wishes to switch on. If a GUI provides a similar feature, a specific button would directly control the light. In the case of spoken interaction the system would have to query the user: "Which light? The ceiling light or the table lamp?"
- The ASDM also performs dialogue tasks. Control tasks, however, must be combined with these dialogue tasks. As a result of the dynamic combination

[1] SOFIA is funded by the European Artemis programme, 2009–2011, http://www.sofia-project.eu

the user is able to utter commands even *during* a conversation. This multitasking capability is one of the main requirements within the IE context. We present our solution to this issue in Sect. 3.6.1.

According to the requirements for developing an ATRACO system that supports several ASs, we have defined as set of dialogue models for different purposes. These models have been implemented and tested as part of the ATRACO social evaluation. The dialogues are presented in Appendix A and the results of the qualitative evaluation are analysed in Sect. 5.5. The provision of specific task-dependent models allows for a dynamic dialogue combination. Thus, the ASDM is able to provide a consistent interface that allows for accessing, for example, more than one device (i.e., control task) via voice in parallel. In Sect. 4.1 we present the architecture of the ASDM and show how it is integrated into the ATRACO IE. In the following we provide an overview on different approaches to HCI within IEs and explain how they influenced our own approach.

2.3 Interaction Within Intelligent Environments

International research projects have been concerned with multimodal spoken language dialogue interaction within IEs for several years. The project SmartKom investigated and tested concepts for the development of new forms of HCI (Berton et al. 2006). The aim was the exploration and development of a self-declared, user-adapted interface for the interaction between humans and devices. The advantages of natural speech communication have been combined with those of graphical and gesture-based user interfaces. The focus was placed on multimodality. The interaction, however, was limited to in-car scenarios and dealt with specific tasks (such as route planning and parking place reservation). Our work is not limited to specific tasks but emphasises the necessity of interfaces that are able to adapt to the current situation and the tasks this situation is shaped by.

A further example pointing into a similar direction is the digital pocket-sized assistant, developed by the Medication Advisor Project (MAP). It offers new technical mobile solutions through the integration of multimodal interactive functions, new assistance systems, agent-based technologies, and multimedia (Ferguson et al. 2002). Again, a domain-specific solution how spoken dialogue may be utilised has been investigated: the societal problem of helping people managing their medication. Contrary to these approaches we developed a generic framework for SDM that can be utilised independently from the task it should realise. EMBASSI allows the homogeneous multimodal remote control of all electronic appliances of everyday life. The aim of this project was to minimise the complexity of user interfaces and operating instructions (Hildebrand and Sá 2000). Here, the authors do not limit the spoken interaction to specific tasks but allow the control of various devices similarly to our approach. However, they are limited to command-and-control dialogue layouts and avoid the introduction of more complex dialogue layouts due to their higher complexity.

The Project D'Homme points into a similar direction: Quesada et al. (2001) developed an SDS that is able to understand natural language commands for controlling an IE. They combined a semantic-oriented grammar with specific dialogue moves that allow for combinations of different user inputs such as "Turn the kitchen light on and the bathroom off". Due to its computational complexity we assume that the grammar the authors proposed is beneficial for commands within a limited scenario. However, it cannot be applied to more complex dialogues for information retrieval or gathering (especially within an evolving IE domain). Within the project SmartWeb methods and technologies were investigated and implemented in the scientific fields of intelligent user interfaces, semantic web, and information extraction (Sonntag et al. 2007). The authors utilise a comprehensive ontological knowledgebase in combination with the rule-oriented Information State approach (see Sect. 2.4.3). This combination allows for advanced question–answering dialogues that are undoubtedly useful—especially within IEs.

However, compared to our approach we do not consider the existence of a predefined knowledgebase consisting of a fix number of individuals. We rather face an emerging environment. Thus, we must integrate new dialogue-related information into the decision logic of the SDM during runtime. The issue of an emerging and changing environment is discussed in Coutaz et al. (2005). In addition to the general approaches to SDS in the IE domain, the term *adaptation* is also a topic of scientific discussion. In the following we present a general definition and discuss alternative approaches to realise adaptation. McTear (2004) distinguishes between different types to adaptation. On one hand, the author describes adaptable and adaptive interfaces. The former allows the user to apply changes to the system's dialogue strategies. The latter defines the automatic response of the system to, for example, issues that may occur during the dialogue. On the other hand, McTear argues that adaptation may also happen during an ongoing dialogue or over time (i.e., before or after the dialogue). In Chap. 3 we will present a more granular definition of adaptation that is derived from the three parties involved in a spoken dialogue: the user, the SDS, and the IE. Different aspects of adaptivity in SDS have been touched in the recent past. In the following we present three approaches dealing with error recovery, user adaptation, and a dialogue concept called over-answering.

The adaptive version of the TOOT dialogue system (Litman and Pan 2002) is able to change the underlying dialogue strategy. It supports (1) an optimistic layout using user initiated dialogues in case there are only few recognition errors detected and (2) a more conservative strategy using, for example, directed questions. As part of the InterAct project, Jokinen et al. (2002) investigated how spoken dialogue can be adapted to different user characteristics. This also affects the output of the SDS—in some cases short answers are more favoured than a polite and comprehensive statement by the system. A further challenge is to handle over-specification or over-answering situations during spoken dialogues. Such situations occur if the user provides more input than the system actually expects to receive. Regarding this, Qu's works about information research with mixed dialogue-initiative is of interest (Qu 2001, 2002). The author uses constraint processing techniques to support initiative-taking actions within a form-based information dialogue system. The VoiceXML dialogue description language also provides possibilities to realise dialogues to handle over-answering situations (see Sect. 2.4.1). Our proposed SDM framework utilises VoiceXML as primary dialogue description language and therefore in principle supports this type of dialogue strategy.

The described state of the art in dialogue modelling and management research differs from our requirements for adaptive spoken dialogue within IEs. We consider the IE to be a provider of emerging knowledge. Thus, we must adapt to the current state of the environment and the situation of use, which is usually implied by the user. Furthermore, smart devices are mostly used by non-specialists and increasingly frequently by disabled persons without particular knowledge of computer equipment and in their usual context of life (Lockwood and Cook 2008). Therefore, such systems must be easy to use, non-intrusive, and must exploit the most natural communication means. Undeniably, enhanced communication and assistive capabilities increase the usability and the social acceptability of smart systems. However, SDSs that need to support a complex and unconstrained dialogue interaction under different conditions of use are still the subject to research. In particular, there is a need to investigate on how to improve the interaction between the user and the environment by endowing SDSs with more intelligence. As a result such systems would not only be able to retrieve information, but also to integrate information from multiple sources. This would allow for resolving potential conflicts and problems that may occur if the user context changes. This is one of the most important topics that we focus in our work. An *assistive* and *adaptive* SDS is a competent and sensitive complex multi-functional technical system, able to perceive and to interact in a complex and dynamically varying environment (Minker et al. 2009). This environment is able to transform perceptions into a model-based internal representation, to acquire information, and finally, to react accordingly, i.e., to generate and to perform actions based on the information at hand.

The integration of contextual information into the decision logic of an SDM has recently been discussed by López-Cózar and Callejas (2010). They focus on the importance of a user model describing a profile that provides, for example, the gender, the mother tongue, and information about the experience with a spoken interface. This information may be utilised by the SDM to derive, for example, a suitable repair strategy. We have evaluated this issue as part of this document in Sect. 5.4. The influence of user profiles on SDMs has also been investigated by Vipperla et al. (2009). The authors compared the error rates of an SDS used by young adults (mean age of 22) with the error rates of the same SDS when it is used by older subjects (mean age of 66). Their results indicated that an ASR component used within the IE context must be adapted to both the domain of usage and to the acoustic and linguistic characteristics of the users. The modular architecture of our approach facilitates the integration of user-related information and allows for a user-centred selection of all SDS components including the selection of a proper dialogue model.

Within the IE domain several international research projects have been concerned with developing SDMs. Two directions have been discussed in the recent past: heavyweight rule-based frameworks such as the TrindiKit (Larsson and Traum 2000) and statistical approaches such as the Bayes Net Prototype implemented within the TALK Project (Young et al. 2006). The former requires strong assumptions regarding the set-up and adjustment. Once the rule-base is implemented, the system performs well. However, for more complex dialogues, the rule-base is getting increasingly complex as well. The latter approach relies on the availability of training data, which appears to be a major disadvantage since it is costly to collect corpora and to train the statistical models. The proposed role of the SDM in the two approaches is explained in Sects. 2.4.2 and 2.4.3. In addition to these general approaches to SDM, the InterAct project developed a specialised IE SDM that adapts to the context (availability of devices/services) and the physical status of the surrounding environment (Montoro et al. 2004). The InterAct IE provides a blackboard, which is used by the dialogue manager to generate a so-called dialogue tree that describes all available grammars, utterances, and system commands. The dialogue tree is used to derive utterances and grammars in order to be able to execute system commands. In terms of service/device control, these ideas are closely related to our approach. The project Gaia (Román et al. 2002) aims to construct an infrastructure for IEs by developing pervasive applications using an agent-based middleware. This middleware uses ontologies to describe the semantics within various contexts. We also include such semantic values in our SDOs but mainly use them to define the flow and status of the ongoing dialogue within the actual context. A multimodal user interface for IEs was presented in Ruser et al. (2003). This system combines a graphical view of a smart home with a spoken command system that adapts on the device level. However, we aim at a more general model of spoken dialogue that is not limited to command-and-control but also provides communicative capabilities such as negotiation. Therefore, it must also adapt to events and to changing tasks.

Recently, several approaches utilised a logical framework underlying the SDM. Gnjatović and Rösner (2008) use the rules of the game Tower-of-Hanoi to process user commands and provide system utterances in a logical consistent manner. Bühler (2009) aims at domain-driven dialogues by modelling the domain using logical expressions. The author also adds user requirements (constraints, assumptions, etc.) that can be processed by a domain-reasoner. This reasoner is able to avoid logical inconsistencies and can be used to solve over-answering situations. These approaches are interesting as long as it is possible to use logical statements to express the domain. However, a domain also includes the goal and the current context that cannot be logically expressed in all cases. Nevertheless, these ideas are relevant to our work since we also aim at a logical consistent model of spoken dialogues that can be utilised to express both structure and state of the dialogue. In the following section we discuss related work in the field of SDM and describe technical details and unique characteristics of the different approaches.

2.4 General Approaches to Spoken Dialogue Management

Since an SDM is responsible for controlling the *content* and the *flow* of a spoken dialogue, it provides a key factor in ensuring a user-friendly and consistent user–system interaction. Several approaches to how and when to select the correct content exist. Early systems used simplistic prescribed state machines combined with grammars. Later on, heavyweight plan-based systems have been implemented and nowadays several stochastic corpora-based prototypes exist. Given the exposed position of an SDM regarding dialogue content and flow, it is not far-fetched to implement complete SDS frameworks defined by the technical structure of the SDM. In this case the acoustic front-end and the semantic layer are usually handled as separate modules that can be substituted if necessary. After discussing the different approaches to SDM, we present possible enhancements of SDMs in Sect. 2.5.

2.4.1 State-Machines and Grammars

The main purpose of an SDM is to control the dialogue between user and system. Like many other computer science disciplines, the early approaches to dialogue control had their roots in the 1940s and the 1950s. As described by Jurafsky and Martin (2000) two paradigms that provided a basis for today's science influenced SDM the most: the automaton (Turing 1937) and the information-theoretic models (Shannon 1948). In this section we focus on the role of the automaton theory. Turing's model of algorithmic computation, which paved the way for the automata theory, enabled the development of finite automata and regular expressions (Kleene 1988). The founder of information theory, Claude Shannon, added probabilistic models to automata so as to describe languages. Seized on this idea, Chomsky utilized finite-state machines to define grammars that determine a finite-state language and provided a categorisation for the description of languages (Chomsky 1956). These approaches funded the formal language theory. Regarding SDSs the context-free grammar plays a major role since most of the language models in nowadays ASR base on this type of grammar. A context-free grammar is defined by the 4-tuple $G = (\sum, V, S, R)$ where:

1. $\sum$ is an alphabet (of all terminal symbols)
2. V is an alphabet of all non-terminal symbols (or variables)
3. $S \in V$ is the start symbol
4. R ia a relation from V to $(V \cup \sum)$ such that $\exists \omega \in (V \cup \sum)* : (S, \omega) \in R$

A more detailed definition may be found in rí Adámek (2008). Context-free grammars have first been defined by Chomsky as part of the Chomsky hierarchy and have independently been discovered by Backus and Naur who established the Backus–Naur Form (BNF) (Knuth 1964). BNF can be utilised as a notation technique for such grammars. All these innovations paved the way for complete parsing systems, such as the Transformations and Discourse Analysis Project (TDAP) by Zelig Harris (Nevin and Johnson 2002). Harris used cascaded Finite State Transductors (FST) in order to understand the user input. The system started with a dictionary lookup. Words that were not part of the dictionary could not be parsed. Afterwards so-called "grammatical idioms" (i.e., for example, per hour, etc.) have been detected before several rule-based disambiguation techniques were assigned. Once this processed cleaned up the user input, the mentioned cascade of FSTs detects "simple noun phrases", "simple adjuncts", "verb clusters", and the clause(s) the input consisted of.

In the past, many approaches to dialogue control, to language modelling, and, in combination, to SDM have been presented. They take advantage of grammars and of state-machines. Nowadays, the de facto standard of dialogue control and definition is the W3C VoiceXML 2.1 specification (Oshry et al. 2007), which is also based on state-machines. The aim of VoiceXML is to transfer the ideas of one of the most fundamental web technologies—namely hypertext mark-up languages (HTML)—to SDS applications. The ease-of-use of the spoken dialogue description language allowed for a wide spread of "voice hosters". These host spoken dialogues in a similar manner web hosters provide Internet content. Listing 2.1 illustrates a VoiceXML application consisting of a question by the system ("Do you want to start?") and a specific grammar describing the possible user answer. In this example the user may answer "yes" or "of course" to indicate acceptance or "no", "never", and "nope" to decline. The question–answer pairing is encapsulated within the *field* tag that allows to connect the specific field with a semantic meaning. Possible meanings are "positive" or "negative" that can be passed to, for example, an application. The VoiceXML standard does not only define how a dialogue has to be described but also specifies an abstract processing model for VoiceXML applications. This model is called the Form Interpretation Algorithm (FIA). This algorithm is responsible for selecting the accessible items within the actual VoiceXML form. Afterwards it either prompts the user and/or waits for a specific user input that must be analysed using the appropriate grammar. If the second phase is successfully terminated, the user input can be processed by filling one or more items that accept corresponding values. In case an exception occurs, proper actions may be initialised by, for example, carrying out a further inquiry.

```
1  <vxml version="2.1">
2   <form>
3    <field name="move">
4     <grammar type="application/x-gsl">
5      [[yes (of course)]{<move "positive">} [no never nope]{<move "
          negative">}]
6     </grammar>
7     <prompt>
8      Do you want to start?
9     </prompt>
10   </field>
11  </form>
12 </vxml>
```

Listing 2.1 A VoiceXMl application

VoiceXML partly extends the straightforward approach of defining a state-machine that represents the dialogue by the metaphor of *frames*. These frames can be used to describe the structure of a dialogue more freely in the sense that specific situations, such as over-answering, can be handled. Nevertheless, the dialogue itself is still represented as a state-machine. In the following we discuss two underlying spoken dialogue strategies that can be realised using VoiceXML:

- The *system-directed* dialogue strategy is the most usual way spoken dialogues are realised nowadays. Here, the system asks the user for specific input and decides which dialogue step to be carried out afterwards. VoiceXML is perfectly suited to define this kind of rigid dialogue management strategy. Several constructs, such as re-prompts, error catch declarations, etc. may be utilised to achieve a coherent question–answer-confirmation layout.
- The *mixed-initiative* layout can also be realised using VoiceXML. However, the classic meaning of mixed-imitative being a mixture of system and user directed dialogue is not fully supported by VoiceXML. Instead, the VoiceXML specification describes mixed-initiative as the capability to handle over-answering situations. Hence, the user may volunteer more information than initially requested by the system. This may lead to more efficient spoken dialogues.

A main lack of VoiceXML is that user-directed strategies are not supported by the standard. However, Schnelle-Walka and Feldes (2009) showed that VoiceXML-based dialogues can be used to develop spoken interfaces within the IE context. The authors introduced a pattern language that fills the gap between command-and-control systems and user-directed interfaces. A main benefit of this dialogue layout is that the user is not restricted to a specific input the system is currently able to understand. However, a disadvantage is that the system requires comprehensive grammars. Otherwise, there is a risk that the user is not aware of the words and the syntax the system accepts. Due to this shortcoming we present an approach to generating VoiceXML snippets representing a specific dialogue turn. In this context we introduce user turns, system turns, and exchanges. This allows for defining spoken dialogues that also support user-directed layouts (see Sect. 3.5.2). The underlying dialogue structure is a further shortcoming of VoiceXML. It only provides *single* task capabilities. Taking the definition of tasks as provided in Sect. 2.2 into account, an ASDM for IEs must handle more than one interactive task in parallel. In VoiceXML dialogue switches between *multiple* tasks have to be defined handcrafted as part of the dialogue definition during design time. We present a solution for this issue in Sect. 3.6.2.1.

2.4.2 Stochastic Approaches

The second paradigm that significantly influenced SDM is the probabilistic or information-theoretic model. It was also introduced by Shannon (1948). His main contributions were the definition of the noisy channel metaphor as well as the decoding theory. Further important bases of the stochastic approaches are Hidden Markov Models (HMMs) and Markov Decision Processes (MDPs). Both Shannon's contributions and the evolutions that started with Markov's theoretical foundation are important not only for communication transmission but also for speech recognition and, as demonstrated in the TALK project (Young et al. 2006), also for dialogue management. HMMs have first been introduced by Baum et al. (1970) and base on MDPs. MDPs are stochastic processes that describe progressions of random variables depending on each other. Unlike MDPs, HMMs do not only provide an end-probability for the entire decision process but also emit a specific output to each state. The model is *hidden* since only the emitted outputs are visible and can be used to reason on the actual sequence of states. Thus, the states themselves cannot be investigated directly. In the recent past a different methodology has gained attention in the field of SDM: the *Partially Observable Markov decision process* (POMDP).

A POMDP is a MDP whose states are partially observable. This is comparable to HMMs but here the states are not totally hidden. Young (2007) presents a methodology to use POMDPs for dialogue management. He argues that a spoken dialogue including its uncertainties can be generally modelled using POMDPs. However, it would be computational intractable to use POMDPs directly since the state space of a practical SDS is usually very large. Therefore, it is not useful without further approximation. Hence, the author proposes to use a partitioning of the states that is based on the Hidden Information State (HIS) approach. HIS has been introduced within the framework of the TALK Project. Stochastic approaches usually outperform comparable state-machine-based approaches. However, all statistical approaches have a main disadvantage in common: they rely on the availability of training data, i.e., of a corpus to be collected in a costly process. Furthermore, the collected data have to be used to train the statistical models (i.e., the POMDPs) so as to provide a specific dialogue system for a specific domain.

2.4.3 Plan- and Information State-based Systems

A further stage in the development of SDSs and more specifically of dialogue control has been the introduction of logic-based systems, which strongly influenced the plan-based approaches. One of the outstanding contributions to this area has been made by Colmenauer by the development of Prolog (Colmerauer and Roussel 1996). A program written in Prolog consists of a database that provides facts and rules. A fact, for example, is an expression such as $mother(sophia,tina)$ implying that Tina is Sophia's mother. Rules in Prolog describe logical implications. They can be defined using the rule-operator $:-$. The following rule can be used to describe that X is the brother of Y:

$$brother(X,Y) :- father(X,Z), father(Y,Z), mother(X,M), mother(Y,M), man(X).$$

Variables can be set using the method of *unification*, which means that all statements are valid as long as the entire expressions can be evaluated to true, if a specific value is assigned to a specific variable. In other words, the above expression can only be evaluated to true if X has the same father as Y, X has the same mother as Y, Z and M are not equal, and X is a man. Utilising such a rule-based approach, dialogues that are inherently logic can be modelled. Thus, practical user–system interaction becomes possible. As an example, we define the following rule base (in natural text, not in Prolog syntax, see Colmerauer and Roussel 1996):

Every psychiatrist is a person.
Every person he analyzes is sick.
Jacques is a psychiatrist in Marseille.

This rule base allows the system to perform a dialogue such as:

Is Jacques a person?—Yes.
Where is Jacques?—In Marseille.
Is Jacques sick?—I don't know.

However, the original rule base necessary to define the dialogue is huge. This implies that the setup and the resulting capabilities impose severe limitations on the practical usage of pure rule-based dialogue systems. An interesting approach to avoid these limitations was presented within the context of the TrindiKit framework. An improvement has been achieved by combining the logic-based Prolog attempt with a Natural Language Understanding (NLU) paradigm: The Total Information State (TIS).

The idea of an Information State goes back to Ginzburg's Dialogue Gameboard (DGB). The DGB describes all information that is needed to proceed with a specific dialogue (Ginzburg and Cooper 2004). A simplified DGB may consist of:

$$\begin{bmatrix} FACTS & set\ of\ facts \\ LATEST-MOVE & (illocutionary)\ fact \\ QUD & set\ of\ questions \end{bmatrix}$$

Here facts refer to shared information between the dialogue participants. It is considered that this information can be taken for granted. The latest move defines the last information exchange that occurred during the conversation. Hence, the latest move can be seen as a pointer to the actual dialogue position. QUD is the abbreviation of *questions under discussion* and consists of a set of currently discussable questions. These questions are partially ordered depending on their conversional precedence. The DGB was refined by Larsson (2002), who introduced the metaphor of *issues under negotiation*. It provides a broader set-up of the DGB, namely the TIS. An important feature of the TIS is that it distinguishes between private and shared knowledge. Thus, it takes the existence of specific knowledge that is unknown to the user into account. The main objective of a TIS-based dialogue system is the transformation of private information to shared knowledge.

Notably, private knowledge modelled within the TIS can be retained by the user and the system. The combination of an Information State and of so-called update rules provides a framework to define dialogues that (depending on the nature of the logical update rules) features generic dialogue behaviours and strategies such as grounding (Clark and Schaefer 1989). The idea of a knowledgebase that describes the actual state of a dialogue and all related information is relevant to our work since we maintain a similar structure within our dialogue models. We broaden the model by adding information about the description (i.e., grammars, utterances, pre-, and postconditions) and the state of the dialogue. In the following section we present different approaches that have been investigated to enhance the capabilities of key aspects of an SDM that behaves adaptively.

2.5 Enhanced Spoken Dialogue Management Methodologies

Enhancements may be applied to an SDM regarding two aspects: the improvement of the understanding capabilities of the SDS and the integration of dialogue strategies. The former refers to the influence an SDM may have on the other SDS components that realise the acoustic frontend and the semantic layer, respectively (see Fig. 2.1). Since the SDM is responsible to provide a coherent dialogue flow, it is also responsible for deciding if a specific utterance has been correctly recognised or not. Furthermore, the SDM has to be able to appropriately react on a presumably wrong recognition. In Sect. 2.5.1 we present several approaches that concern this matter.

The latter enhancement refers to the integration of specific dialogue strategies into the dialogue flow the SDM may provide. In our work we focus on two kinds of dialogue strategies: topic switching and repair. Especially within the IE domain where the SDM must establish a spoken interface for multiple tasks that may run in parallel, it is necessary to integrate topic switching strategies into the dialogue generation. In Sect. 2.5.2 we present related work concerning topic switching. So as to render the dialogue coherently it is also necessary to include repair strategies. Therefore, we have investigated how different user groups rated specific repair strategies. In Sect. 2.5.2 we present the relevant work within this area.

2.5.1 Understanding Methods

In addition to the general SDM methodologies presented in Sect. 2.4, adaptivity can also describe sophisticated behaviour of the dialogue control itself. Within this context many approaches relate to adaptive understanding, i.e., adaptive interpretation of spoken input. The reason for this seems to be obvious: a prerequisite to allow the SDM to decide how a dialogue should continue is to correctly understand the user input. However, since speech is not a "crisp" communication channel, difficulties arise when a computer interprets spoken user input (McTear et al. 2005). Such difficulties usually can be ascribed to misinterpretations caused by the recogniser, for example, when it is not possible to map an audio signal to a word that is part of the applied grammar.

One possibility to decide if an input can or cannot be mapped is to calculate a "confidence measure" (Jiang 2005). It is usually a value between 0 and 1. To detect if a successful mapping can be assumed, the confidence measure has to exceed a predefined threshold. The process of recognising an input that cannot be successfully mapped to a word defined in the grammar is referred to as "non-understanding". In contrast, and independent from a contextual and semantic correctness, the successful mapping of an input to a word is referred to as "match". It is beneficial to avoid the occurrence of non-understandings as far as possible. However, solving this issue by implementing huge grammars that cover nearly all possible inputs would not be beneficial. Huge grammars may also lead to more misunderstandings. These may even be more destructive to a dialogue than a non-understanding would be. To avoid misunderstandings, the grammar should be kept as small as possible. However, this would also avoid the SDM to provide intuitive spoken dialogues. Chung et al. (2004) show an approach to stepwise broaden the grammar during a second or third phase of recognition. Here, the grammar is extended depending on the context. This approach is relevant to our work since the ontologies we use as spoken dialogue models are perfectly suited to be extended during runtime.

López-Cózar and Callejas (2006) propose a different approach: the authors apply a second level of recognition. The first level comprises of a comprehensive grammar covering all possible user inputs within the application domain. As an output the first recognition level provides a graph of words. It is a network constituted of words (corresponding to the nodes) and probability transitions between the words (corresponding to the arcs). The second level of recognition comprises an analysis of this graph of words. Three parameters are important for the analysis: a set of word classes (consisting of keywords), the current prompt the SDS uttered before, and the transition probabilities. The authors showed that their approach significantly enhanced the recognition accuracy compared to a similar SDS that utilises a prompt-dependent grammar. A limitation of the proposed technique is that the recognition enhancement does not involve the SDM decision logic. This means, for example, that the system is not able to filter out utterances that do not have a (semantic) meaning within the domain. The approach is relevant to our work since we also apply a second level of recognition. However, we leave it to the decision logic of the SDM if it is necessary to analyse the input of the user twice (see Sect. 4.2.3.4).

Another relevant approach to our work has been presented by Nakano et al. (1999). The proposed methodology called ISSS (Incremental Significant utterance Sequence Search) follows the idea of stepwise analysing the user input without knowing the entire input from the beginning. The authors try to recognise the input on a word-by-word basis and built up a knowledgebase consisting of several possible input variations. This knowledgebase is actualised for each newly recognised word.

Once an end of input has been detected, depending on a probability score, the most appropriate system reaction is selected and provided to the user. In order to modify the knowledgebase the authors introduce the concept of Significant Utterances (SUs). An SU is a user input that is necessary to continue the dialogue. SUs may also consist of more than one SU, which is crucial to allow the word-by-word approach. We also apply the concept of SUs. However, we do not use them for analysing the input on a word-by-word basis but to decide whether the user input is relevant to the dialogue domain or not. In addition to the enhancements regarding the understanding capabilities of an SDM, we have investigated the benefits and drawbacks of dialogue strategies that are presented in the following section.

2.5.2 Spoken Dialogue Management Strategies

Two major types of spoken dialogue strategies are relevant to our work: *topic switching* and *repair*. Several approaches to task or topic switching in spoken human–human and human–computer dialogues have been investigated. In this section we present the most common strategies that guided the implementation of the dialogues that have been used for our evaluation series. Franke et al. (2002) propose mechanisms to successfully handle interruptions during spoken dialogues so as to reduce their disruptive effects. Furthermore, these mechanisms should improve the human performance during a multitasking situation. The proposed interruption techniques have been applied to an SDS deployed for managing military logistics tasks. The standard manner to switch between various tasks is to interrupt an ongoing task and to insert new tasks that may be more urgent. Such system behaviour may negatively affect the human information management capacity. To avoid those distracting effects of dialogue interruption, the authors aimed at furthering efficient task resuming.

In their approach the authors divide the user interface design for handling interruptions into three phases: the pre-interruption, the mid-interruption, and the post-interruption phase. The former phase warns the user about switching from the main task to the interrupting task. The assistance may support the user to distinguish the main task from the interrupting one. In a case study the authors utilised a different voice for the interrupting task to warn the user. The mid-interruption phase is focused on the user's transition to the interrupting task and the user's ability to maintain situational awareness of background tasks. To implement the mid-interruption the authors rely on McFarlane's taxonomy of human interruption (McFarlane 2002). McFarlane defines four primary methods for coordinating human interruption: (1) immediate, (2) negotiated, (3) mediate, and (4) scheduled interruption. Which of these interruption solutions should be applied to the application depends on the relative importance of the alerting task related to the current task. If the interrupting task is critically important compared to the ongoing task, the interruption occurs immediately. If it is less important it occurs case-scheduled. In all other situations the interruption is negotiated on a per-case basis as this was measured to be the most promising strategy. The post-interruption phase provides task recovery to the original task that was interrupted. In their test application a user request provided information about the previous task. To test their approach to efficient interruption assistance the described mechanisms were implemented into the Listen, Communicate, Show (LCS) system in which mixed-initiative spoken dialogue interaction has been integrated (Daniels 2000).

Human–human Multitasking Dialogues (MTD) characterised by speakers who switch from a main task to an alerting task have been studied in Yang et al. (2008). In this work the authors investigated guidelines for building human–computer spoken dialogue interfaces that support multitasking. They concentrated on finding definitions when and how to switch between tasks. They discovered that the dialogue partners usually use so-called discourse markers and prosody cues to signal task switching. Furthermore, the subjects tried to switch when they assumed it would be less disruptive. To foster the understanding of human–human MTDs, the authors collected the Heeman's MTD corpus in which dialogue partner pairs perform overlapping verbal tasks (Yang et al. 2011). To collect this corpus the following dialogue situation has been defined: A poker Gameboard forms the main task. The users are playing together and communicate via voice. The game interface is displayed on a screen. The (probably interrupting) real-time task is formed by a second game where the players should find out whether the opponent has a certain picture on the bottom of his display. Two flashing bars were used to alert the players. The authors analysed the collected data in order to figure out how the subjects switch between tasks. They discovered that the players strive to switch to an alerting task at a less disruptive point during the ongoing task namely at the end of a round or at the end of the game.

Furthermore, the authors examined how the users signalised the task switching. The players often used discourse markers such as "oh" and "wait" while switching from the ongoing task to the interrupting, real-time task. The prosody of the phrase "Do you have" during the current task and by task switching was studied as well. For this purpose the pitch, the duration, and the energy were measured. Relying on the test results Yang et al. established that higher pitch is related to task switching, but the energy and the duration are merely irrelevant. Additionally they demonstrated that the pitch correlates to the situation of switching: the higher the pitch, the more disruptive the interruption. Finally, machine learning experiments have been conducted to determine if the described cues are reliable for discriminating task switching. For this purpose a decision tree classifier was used. The results of the experiment showed that discourse context and normalised pitch are useful features for task switching recognition.

The second important dialogue strategy type is related to error recovery. In general there are two types of error recovery: explicit or implicit. An example of implicit error recovery is proposed by López-Cózar and Callejas (2008). The authors developed a post-correction technique to repair errors before the SDM incorporates wrong knowledge into the dialogue. In our work we focus on explicit error recovery. We assume a specific input has been wrongly recognised and the SDM must repair this failure. Bohus and Rudnicky (2005) gave an accurate overview on this topic.

The authors investigated how different repair strategies compare to each other in terms of recovery rate when facing a non-understanding error. Ten different strategies were examined, including the so-called MoveOn strategy. Here, the system proceeds with the dialogue after an error occurred in the hope of collecting ancillary information that can be used to repair the failure. This method can be classified as semi-implicit, since the user does not need to explicitly infer in all cases. In their work MoveOn was the repair strategy with the highest recovery rate, followed by a full help strategy, a concise "You Can Say" and a straightforward re-prompt strategy. The second part of this work was an analysis of the efficiency of a repair policy. Taking a metric into account, which chooses the expected best strategy in every step, the task success rate significantly increased. This demonstrates the potential of improvements, when adaptively choosing the most promising strategy in a specific situation.

In this context, the MoveOn strategy by Skantze (2003) is also of interest. Within a routing domain the author tried to ignore a possible uncertainty of a recognised utterance and instead asked a different (but topic related) question. This also led to a significant improvement of the recovery rate. However, whether and how this concept can be adapted for other domains has still to be clarified. A main reason for the effectiveness of this concept is that no misunderstanding or non-understanding is signalled. Other studies showed that once such an error occurred (even if it was successfully resolved), the prosody and pronunciation changed in a way that the ASR has more problems to recognise the user input (Swerts et al. 2000). This indicates that the user's trust in a proper working system could avoid this disturbing factor. In our approach to ASDM and for the development of the prototype we have developed specific methods for both *topic switching* and *repair strategies*. Both types have been evaluated regarding their efficiency and user-friendliness. In the following section we provide the essentials in brief about our approach before we detail the methodologies and functionalities of our work in Chap. 3.

2.6 Conclusion

In this document a novel approach to Adaptive Spoken Dialogue Management within Intelligent Environments is presented. The major characteristic of our proposed methodologies in this area is *adaptivity*. The prototypic ASDM utilises domain-dependent dialogue models that define both the states *and* the structures of spoken dialogues. We combine Information State-based methods (describing the dialogue state) with state machine-based techniques (describing the dialogue structure). In contrast, established SDM frameworks formally define either the state *or* the structure. Our novel attempt to adaptive spoken dialogues description offers benefits especially when the SDS must provide a coherent interface to a changing application domain. State-of-the-art SDSs provide interfaces to specific, prescribed applications or to a fixed set of devices and services. As opposed to this we developed a framework that allows for incorporating spoken dialogue models for a *varying* set of applications. Our techniques base on the idea of an environment surrounding the user like a digital bubble (cf. Beslay and Hakala 2007). Since it emerges during runtime, the spoken interface must emerge accordingly.

The dialogue logic is strictly separated from the domain-specific information and therefore guarantees a maximum of flexibility and integratability. The ASDM generates spoken dialogue descriptions depending on the models that are currently activated. Since the models are directly influenced by the context that in turn is provided by the IE, the ASDM is able to adapt to the context of the environment. An ASDM is termed to be *adaptive* since it adapts the dialogue according to the following stakeholders: the user, who demands for an efficient and user-friendly interface, the IE that is subject to an ongoing change regarding the availability of devices and services, and the SDS, which may be error-prone. In Chap. 3 we present seven levels of adaptation that correspond to the three parties involved and define the most important challenges our research faces. Our techniques differ from the presented methods for SDM since we do not only focus on dialogue-inherent mechanism but also incorporate adaptivity into the presented ASDM framework. In the following section our approach is presented in detail and the functioning of the prototype is discussed.

Chapter 3
Novel Approach to Spoken Dialogue Management in Intelligent Environments

Intelligent Environments consist of various entities and in parallel provide and execute different tasks. Since neither the sets of entities nor the tasks may remain constant while the user interacts with the system, we speak of a *changing nature* of such environments. Hereunder we have revealed *adaptation* as the major characteristic of an SDM allowing for a consistent interface provision. We have presented several approaches to SDM that partly cover specific aspects of adaptation or adaptivity in Sect. 2.4. However, in order to develop a system that provides *adaptive* spoken dialogue within IEs, it is necessary to denote a general definition of *adaptation*. This definition concerns the three main stakeholders involved in spoken interaction: the user(s), the SDS, and the IE. Of course, the fourth party involved is the ASDM, which seems to play a key role: while the ASDM must handle adaptation, the other parties provoke it. In the following we discuss our proposed definition and provide a complete description of adaptivity regarding the stakeholders mentioned above.

In Sect. 3.1 we define how to adapt to specific, usually non-verbal information about the user. This information may encode his emotional state and behavioural parameters, such as "being in a hurry". It is usually fraught with uncertainty. Therefore it cannot be integrated easily into the dialogue models of the ASDM. However, for a proper and adaptive reaction to the user input these data cannot be left aside. It may be provided by an external "emotion recogniser" or a similar component that is not part of the ASDM itself. In the following we refer to this as *User-centred Adaptation*. A second reason for adaptation is the SDS itself. Depending on the quality of recognition and on the performance of the system, the ASDM must be able to modify dialogue strategies and to choose the optimal methods of recognition. In Sect. 3.2 we discuss different ways to enhance the understanding performance of the SDS. This may be done by either adapting strategies used for error recovery and error prevention during task switching or by adapting the recognition method of the SDS. An example of recognition adaptation is to use, depending on the actual dialogue situation, either grammar-based or free-text recognition (see Sect. 4.2.3.4).

T. Heinroth and W. Minker, *Introducing Spoken Dialogue Systems into Intelligent Environments*, DOI 10.1007/978-1-4614-5383-3_3,

In the following we refer to this aspect as *SDS-centred Adaptation*. Several challenges arise from the role of the ASDM to provide a ubiquitous interface and the aspiration to react appropriately to the various reasons for adaptation. One of the main issues is to enable the system to provide multitasking capabilities. This is an essential precondition to render spoken dialogues adaptive. We assume that a use case describing a common situation within the IE context would usually consist of more than one task (e.g. controlling lights *and* prepare a dinner). Nevertheless, the ASDM must provide a dialogue that handles all tasks the current use case consists of. Compared to GUI-based multitasking, a main difference here is that the ASDM must associate the utterances of the user with the appropriate task. A GUI would, for example, provide several widgets for the specific tasks allowing for a "crisp" input by clicking on a dedicated button. Hence, we assume the third reason for adaptation is the IE. In Sect. 3.3 we define this aspect, called *Environment-centred Adaptation* in detail. An application scenario illustrating the issues the ASDM should handle is presented in Sect. 3.5. In Sect. 3.5.1 we extract the episodes underlying the scenario and map them to our proposed definition of adaptation. In Sect. 3.5.2 we describe the fundamental conversational acts that are used to define the dialogue model underlying the developed framework. Building upon that, we present the functionalities that render the ASDM adaptive in Sect. 3.6 and relate them to the specific episodes extracted from Sect. 3.5.1. The general definition of adaptation we present here bases on the separation of the stakeholders. This separation implies the provision of an ubiquitous interface to a changing context (Abowd et al. 1998). Figure 3.1 illustrates the four parties and their relationships.

This classification is in accordance with the term *adaptivity* as defined in Minker et al. (2009). The authors outline this to be one of the most important characteristics of SDM acting meaningfully within changing domains. However, it is not completely defined what adaptation means in this context. For a definition of adaptation it is not sufficient to only postulate *what* has to be adapted. To a greater degree it is necessary to define *why* the adaptation occurs. The main aim is to adapt the *interface* between the user(s) and the computer-based system, whatever that may be like. In this work we focus on the spoken dialogue interface that adapts to the status of the stakeholders introduced above. These stakeholders also answer the question *why* the interface must adapt:

- Without any doubt the *user* is the main reason for the existence of a speech interface. However, not in all cases he is the main reason for adaptation. An expert user would be able to control an interface that does not adapt to anything beyond the input that has been recognised. In most cases a reason for adapting a dialogue with respect to the input of the users is to provide a more comfortable interface. Such an interface may also be used by novices. In the following we present two variations of adaptation that are caused by the user: *Emotional Adaption* and *Behavioural Adaptation*.
- The *Spoken Dialogue System* primarily consists of modules for speech recognition and synthesis. Since in practice the ASR does not provide a perfect recognition rate, an ASDM should adapt to errors of various types. In case the recogniser is not able to provide an appropriate result, the grammar and

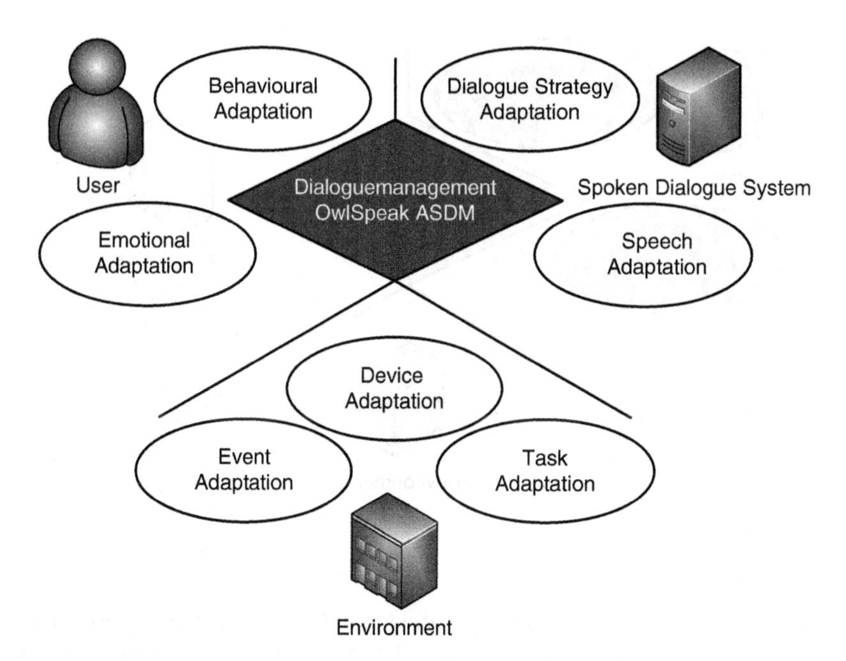

Fig. 3.1 The main factors that influence the ASDM

the language model can be adapted by the ASDM in order to re-analyse the spoken input. The ASDM may also influence the speech synthesis in case the user does not understand the system properly. We call this type of adaption *Speech Adaptation*. Besides influencing the SDS by adapting the information on the "word level", it is also possible to adapt on the "dialogue level". For example, if error recovery or error prevention becomes necessary, the ASDM must modify the dialogue strategy. Here we speak of *Dialogue Strategy Adaptation*.

- The *Intelligent Environment* is the third main stakeholder. It consists of devices and services that are interconnected with each other. In Heinroth et al. (2010) we have defined three levels of adaptation that are caused by such a changing environment: *Device Adaptation*, *Event Adaptation*, and *Task Adaptation*.

Before we present a scenario to illustrate our approach and to discuss dialogue situations that require adaptation, we focus on the seven fundamental adaptation levels caused by the three stakeholders.

3.1 User-centred Adaptation

As depicted in Fig. 3.2 we define User-centred Adaptation to consist of the two levels *Emotional* and *Behavioural Adaptation*. These levels must be handled by the ASDM. In the recent past several approaches to detect and to model the emotional

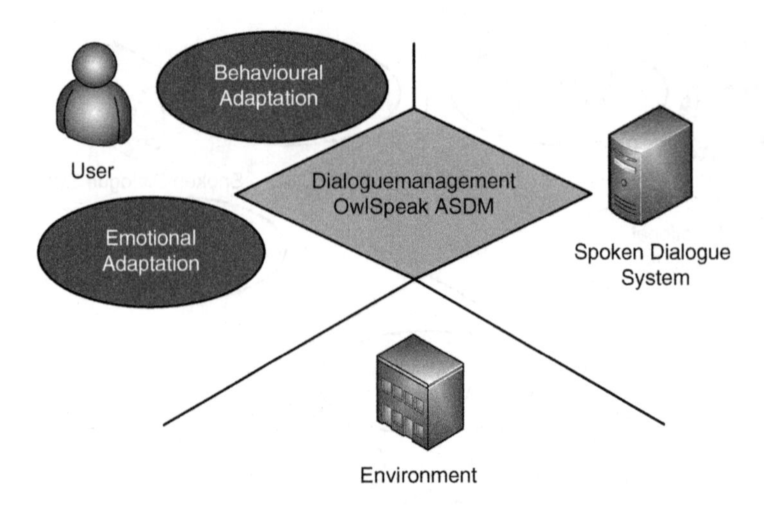

Fig. 3.2 User-centred Adaptation

level of adaptation have been investigated. One of the most complex parts of this level is to correctly model the user state. As part of our work we do not focus on emotional user modelling. However, for the sake of completeness we present two theories of emotion categorisation that may be utilised to influence the dialogue generation. An approach to the categorisation of basic emotions has been introduced by Cornelius: the "Big Six" (Cornelius 1996). The author proposes a set comprising happiness, anger, disgust, sadness, surprise, and fear. This set may be used to describe the emotional state of a human being. A more granular characterisation has been made by Plutchnik: the "Emotional Wheel". Figure 3.3 depicts the eight classes the author proposes (Plutchik 1980; Cowie et al. 2001). The author avoids the use of happiness. Instead he defines acceptance, joy, and anticipation as more fine-grain positive emotions. These fundamental emotions are arranged on a circle. Thus, the angular measure is used to define a broad set of emotions, e.g. optimism between joy and anticipation (67,5 °) and disapproval between sadness and surprise (247,5 °). For the incorporation of emotions into an SDS it is usually more convenient to define a set of basic emotions (Pittermann et al. 2009). Otherwise the emotion recognition process, which is a problem in itself, would become intractable. In Schmitt et al. (2009) we have presented an approach to emotion-aware VoiceXML applications. The results of this work have been taken into account during the decision process of choosing OWL ontologies to define the dialogue model of the ASDM. The open and flexible model easily allows including the data we have introduced for the emotional adaption of the dialogue.

The second level of user-centred adaptation is *Behavioural Adaptation*. Here the spoken dialogue has to react to specific user behaviour. The term *user behaviour* is relatively vague since in the broadest sense it may be everything that is caused by the user and everything the user is involved in. Dretske (1991) tries to detail the term

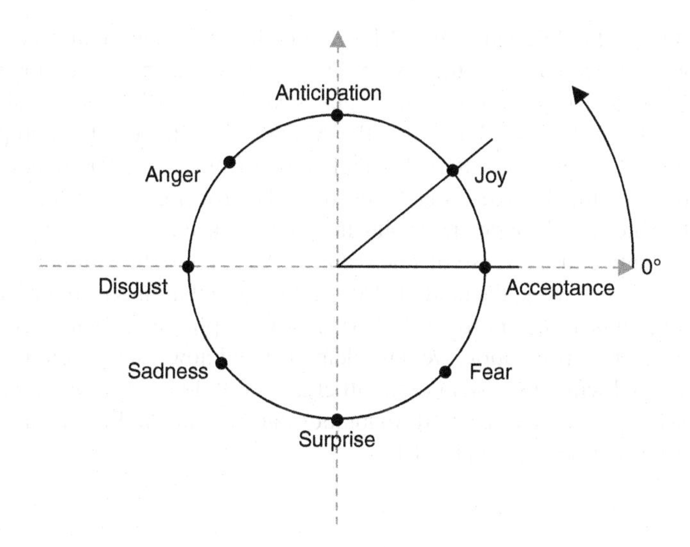

Fig. 3.3 The Emotion Wheel (Plutchik 1980; Cowie et al. 2001)

behaviour. He distinguishes between internal and external causes of behaviour and between goal-intended and goal-directed behaviour. As stated by Warren (2006) herewith the internal representation of the organisation of the behaviour is not sufficiently defined. A main lack is the mere correlation of internal and external facts. If we take more comprehensive explanations into account it seems to be obvious that an SDS would have to incorporate an unlimited number of behavioural variations. This would lead to an intractable model and would thus be comparable to the issue of classifying emotions. Therefore, we define *Behavioural Adaptation* as the characteristic of the ASDM to react to *dialogue-related* user behaviour. Thus, a dialogue model must provide data necessary to react to situations where the user, for example, is in a hurry. Such behaviour may be detected by an external recogniser feeding the ASDM with ancillary information. In addition to such non-verbal data, verbal information expressing the behavioural state of the user should also be part of the ASDM's knowledge base. Utterances such as "Don't bother me", "Skip this", or "Go into details" imply a specific behaviour of the user that apparently should influence an ongoing dialogue.

As illustrated in Fig. 3.2 we differentiate between *Emotional* and *Behavioural Adaptation* since both paradigms lead to different mechanisms the ASDM must implement. For instance, the former one would lead to a dialogue strategy praising the user who is in a negative emotional state so as to reconstruct his motivation. The latter paradigm would lead to more pragmatic dialogue generation decisions such as cutting a long story short and in extreme cases terminating the dialogue. In practice, the reasons (i.e. the triggers) for both types of adaptation may be recognised by external modules. In fact, the ASDM is not able to recognise them in the first place. The external modules that may be realised as online monitors receive the input of the recogniser. They may generate a list of features that can be interpreted by the ASDM (cf. Fig. 2.1). Taking this information into account the ASDM may apply a dialogue strategy and may carry out a specific dialogue step. In case the triggers for *Emotional* and *Behavioural Adaptation* are linguistically encoded, i.e. the user utters "I am sad", the ASDM would be able to appropriately react without the input of an external module (apart from the ASR). To allow such system behaviour, the dialogue model must describe proper grammars or keywords that complete the "core" model describing the dialogue itself. This is one possibility to become adaptive in the sense of User-centred Adaptation. Even though it is not the main topic of our work, the field of User-centred Adaptation is an important one. However, since it is necessary for an ASDM and for the presented dialogue model to handle user-centric information, we have demonstrated how an emotion recogniser can be combined with our system (Schmitt et al. 2009). Further information about the addressed topics of User-centred Adaptation can be found in Pittermann (2008), Bezold (2011), and Schmitt et al. (2011).

3.2 SDS-centred Adaptation

The second main stakeholder is the SDS. Figure 3.1 depicts *Dialogue Strategy Adaptation* and *Speech Adaptation* to be part of what we define as SDS-centred Adaptation. In practice, the most important issue that prevents SDSs to be as wide spread as keyboard- and mouse-based interaction is that speech is not a "crisp" medium. Therefore SDSs are error-prone. The user faces various challenges when using an SDS for managing a specific task:

- A user must be aware of what the system is able to understand. Especially non-expert users want to *intuitively* use a voice interface. This aspect will be investigated further in Sect. 4.2.3.4.
- The system must distinguish between relevant input on the part of the user and negligible utterances. In particular, this may lead to:
  - Misunderstandings (i.e. false-positives) on the part of the system
  - Non-understandings on the part of the system
- The recogniser may provide wrong results or no result at all. This may lead to similar problems as described in the previous bullet.
- The synthesiser may provide output that cannot be understood by the user. This may lead to:
  - Misunderstandings (i.e. false-positives) on the part of the user
  - Non-understandings on the part of the user

Black et al. (2011) present a comparison of three real-life SDSs that perform the same task (telephone-based bus information system) in terms of dialogue word error rate (WER) and task completion. The metric WER is commonly defined as $WER = \frac{S+D+I}{N}$, with S being the number of substitutions, D the number of deletions, I the number of insertions, and N the number of words in the reference. The compared

Table 3.1 Average dialogue word error rate and task completion for the live and the control tests (cf. Black et al. 2011)

	SYS1 (%)	SYS2 (%)	SYS3 (%)
WER			
Control	38.4	27.9	27.5
Live	43.8	42.5	35.7
Task Completion			
Control	64.9	89.4	74.6
Live	60.3	64.6	71.9

SDSs used different techniques including agent-based, VoiceXML, and statistical methods. Besides the main outcome that SDSs in controlled test set-ups perform better than similar systems under live conditions, the authors report an average WER between 35.7% and 43.8%, while a low WER relates to a high task completion rate. Table 3.1 shows the WER and the task completion rate of the compared systems. In Sect. 5.5 the results of the ATRACO-ASDM live tests are presented. In the version of the ASDM used for these tests, the SDS-centred Adaptation has not been taken into consideration. This has led to a suboptimal system performance that is comparable to the results shown in Table 3.1. Taking these relatively high numbers for WER and relatively low numbers for task completion into account, it seems to be obvious that a component is necessary that observes the SDS and selects proper reactions to the various types of errors. We propose the ASDM to be such a component. Its main task is to oversee the dialogue and render a meaningful conversation. We furthermore propose the ASDM to act as an independent device that triggers the ASR and the TTS in order to perform specific tasks. By distinguishing the SDS-centred Adaptation into *Dialogue Strategy Adaptation* and *Speech Adaptation*, an even more fine-grain categorisation can be revealed. The former aspect of adaptation almost exclusively leads to collaborative user–system methodologies to avoid or to repair errors. The latter aspect of adaptation leads to system-inherent enhancements that may only require the user to approve a specific result the system has understood.

We define *Dialogue Strategy Adaptation* to be the ability of the ASDM to integrate specific strategies into the ongoing dialogue to optimise the dialogue flow (cf. Fig. 3.4). Of course, this type of adaptation may also be influenced by Behavioural or Emotional Adaption as described in Sect. 3.1. Several use cases that may occur when using an SDS can be handled if the ASDM behaves adaptively regarding the dialogue strategy. Common examples that closely correlate with the issues discussed above are error recovery and so-called repair strategies. In case the system is able to detect an error these strategies allow for solving an issues such as mis- and non-understandings. The most common repair strategy is a "re-prompt". Subsequent to each dialogue step the recognised utterance is implicitly confirmed. In case of a non-understanding the system re-prompts the question until a valid input is provided by the user. In case of a misunderstanding the user must explicitly interfere with signal words such as "wrong" or "false". This triggers the system to inform the user that the desired correction will be attempted, followed by the simple re-prompt of the question. In Sect. 5.4 we present a comparison of different repair strategies that have been applied to the ASDM (cf. Zgorzelski et al. 2010).

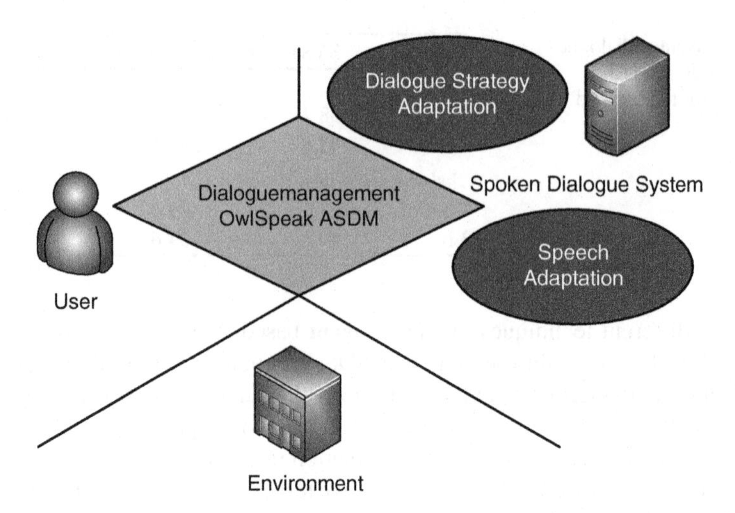

Fig. 3.4 SDS-centred Adaptation

A further category of dialogue strategies are error prevention techniques. If an SDS provides more complex dialogues (combining multiple tasks or processing over-answered input) error prevention strategies may be applied. The main issue that occur when multiple tasks are provided by the SDS is the process of switching between them. When dealing with multiple tasks the user is not only burdened with a higher cognitive load but the SDS also is more error-prone. The system must react to utterances that belong to the previous task in case the user misses the task switch at all. In Sect. 5.3 we present details about a full-scale evaluation of different task switching strategies together with an analysis of their benefits and drawbacks. *Speech Adaptation* is the second aspect defined as part of the SDS-centred Adaptation. It defines the ability of the ASDM to trigger the ASR and/or the TTS to carry out the tasks of speech recognition or synthesis in a modified (i.e. adapted) manner (cf. Fig. 3.4). In this work we focus on adaptation regarding the ASR. Therefore, the system must record the user utterance. If the system is not able to understand the user at the first attempt, this recording can be used by a re-adjusted or a different ASR (e.g. a keyword spotter instead of default grammar-based recognition). A straightforward approach to *Speech Adaptation* is a state-based SDS using prescribed grammars. This SDS utilises an additional keyword spotting technique in case the ASR grammar does not provide a match when interpreting the user input. Table 3.2 shows a dialogue situation that is resolved by applying a keyword-spotter subsequently after the simplified grammar has failed.

In Sect. 4.2.3.4 we present several approaches to such a system behaviour that we call "n-step understanding". During the evaluation sessions we discovered that a dialogue manager that acts adaptively to the SDS significantly reduces the number of non-understandings. It therefore improves the performance of the entire system.

Table 3.2 Dialogue situation with a grammar failure that is fixed by a keyword spotter

Speaker	Utterance	Grammar →	Result	Keyword →	Result
User	The light, switch it on!	–	–	–	–
System	–	[(light on)] →	nomatch	–	–
System	–	–	–	light; on →	The light, switch it **on**
System	Do you want the light on?	–	–	–	–
User	Yes!	–	–	–	–

Summarising, we argue that in order to enhance an SDS in general and, more specifically, to foster the user acceptance, it seems to be not sufficient to only improve the ASR or a module providing linguistic and semantic analysis. It may also be necessary to include the ASDM in the process of *understanding* the user input. Since a major part of *understanding* is the way speech is being recognised, we focussed on this topic and investigated enhanced methods that can be used to combine the ASR and the ASDM. Of course, besides improving the task completion rate by reducing the number of non-understandings, a further stakeholder cannot be left aside: in the following we focus on the IE, which, within traditional SDS frameworks, usually is defined as the back-end (i.e. as the application). In our approach it rather surrounds and therefore integrates the ASDM.

3.3 Environment-centred Adaptation

Contrary to the scheme of a conventional SDS as depicted in Fig. 2.1, our approach does not aim at a system designed to provide an interface to a specific, predefined application. Instead, within the context of IEs, we face a varying and dynamic set of devices, services, applications, and tasks that require spoken interaction. This set of entities changes during runtime depending on the current situation and task. Obviously, in comparison to an SDS that is designed, for example, to perform a dinner preparation task, especially the ASDM has to provide innovative functionality that is discussed in this section. In order to describe these functionalities we have identified three levels of adaptation amplifying Environment-centred Adaptation. The behaviour of the ASDM must be adaptive regarding different types of changes that may happen during (spoken) human–computer interaction in the context of IEs which are explained in the following.

Environmental changes that relate to *Device Adaptation* refer to the continuous modifications of devices and services that are active and accessible within the context of the user (see Fig. 3.5). Depending on the surrounding and the location of the user (kitchen, living room, car, etc.), the availability of devices and services may vary. We assume that users would usually "talk" to devices (i.e. control them via

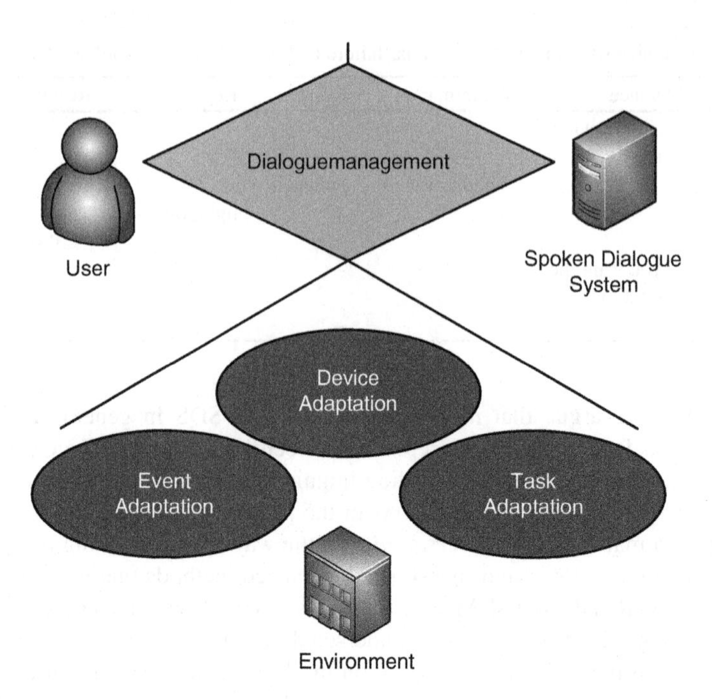

Fig. 3.5 Environment-centred Adaptation

speech) that are within sight and/or that are necessary to achieve a specific user goal. In Fig. 3.6 we present an interactive situation where the IE is aware of the user's intention to find a suitable recipe for the dinner he wants to prepare. The system is also aware that the user usually wants to adjust the light via speech when he arrives back home. The user intention is encoded as a *goal* within a specific task definition that is part of the IE. In the following we refer to this goal as "Dinner Preparation". For the fulfilment of some goals it may be necessary to initiate a spoken dialogue to gain information for the task execution. A trivial example for such a dialogue could be the system asking if a specific task can be started or not. To fulfil the goal "Dinner Preparation" a set of spoken dialogues is necessary: a dialogue that provides information about available foods is required to select an appropriate recipe. A dialogue to control the lights allows the user to adjust the level of lighting. In addition to such controlling dialogues, the most important dialogue is the recipe selection dialogue itself. As depicted in Fig. 3.6 the intelligent bookcase provides the description of a dialogue that is able to answer questions about the availability of books. In our example the user is able to ask for available cookbooks to select an appropriate recipe. Depending on what devices and services are active (i.e. depending on the task definition), the ASDM adds the various dialogue descriptions to its centralised dialogue knowledgebase. We will see that for this purpose the use of ontologies (i.e. the use of Spoken Dialogue Ontologies) bears some outstanding benefits.

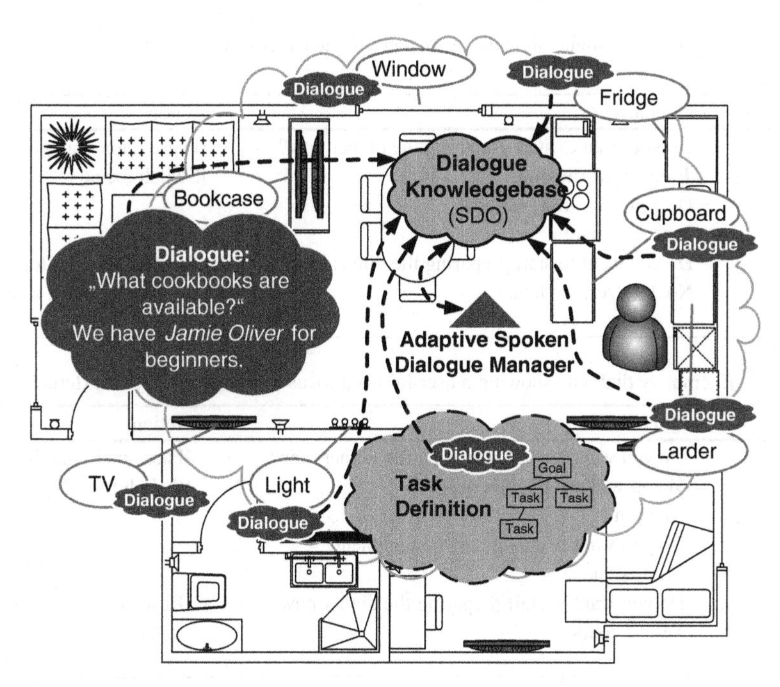

Fig. 3.6 An exemplary set of dialogues that must be handled by the ASDM to fulfil the goal "Prepare dinner for friends" within an IE

Since the dialogue knowledgebase alters during the IE provides a specific goal, the capability to continuously change grammars, utterances, and system commands is required. These dialogue data depend, for example, on a changing entity population and on the changing user location within the IE. Furthermore, it does not seem to be possible to always be aware of the user's activities and the tasks to be carried out (e.g. preparing dinner, doing housework, relaxing, booking a flight). Therefore, it is required to provide the possibility to integrate the control of devices and services that are even not within the current user context but are meaningful from the user's point of view. Thus, the context of the user may also change depending on the time of day and even on the actual level of trust in nearby entities (e.g. guests without access to private information, newly added technical devices). For further information regarding trust and privacy we refer to Konings and Schaub (2011) and Könings et al. (2011). It seems to be obvious that therefore the spoken commands and/or utterances that can be understood or that are uttered by the ASDM will also change continuously. In the case of spoken dialogues an important influence on the effectivity of the communication is the level of noise. In practice, for example, without enhanced audio filters, it will hardly be possible to control an IE via speech while music is playing loudly. However, these types of destructive contextual influence are not within the focus of this work. We assume that in principle (i.e. under changing conditions) it is possible to talk to the SDS and to communicate with the ASDM.

Table 3.3 Exemplary dialogue showing a system-initiated focus switch caused by an external event

Speaker	Utterance	Action
Julia	Do you want to start preparing the dinner now?	[Pause main dialogue]
Julia	**James is at the door.**	[Switch focus]
	Do you want to let him in?	
Suki	Yes, of course!	
Julia	Do you want to start preparing the dinner now?	[Resume main dialogue]
Suki	No, let's resume later	[Set internal event trigger]
	...	

Table 3.4 Exemplary dialogue showing a user-initiated focus switch caused by an external event

Speaker	Utterance	Action
Julia	Do you want to start preparing the dinner now?	[Pause main dialogue]
Suki	**When will my friends arrive?**	[Switch focus]
Julia	At 8 pm.	
	Do you want to change the arrangement?	
Suki	No, it's fine!	
Julia	Do you want to start preparing the dinner now?	[Resume main dialogue]
Suki	Yes, of course!	
	...	

The second level of Environment-centred Adaptation is called *Event Adaptation* (see Fig. 3.5). Since various tasks within IEs are usually to be accomplished in parallel it is necessary to move the actual focus of an ongoing dialogue to other (contingently more urgent) dialogues. These may consist of informative system utterances, alerts, or short yes–no questions. Afterwards, the ongoing dialogue can be resumed. We have recognised two types of events that may imply a focus switch: *external* and *internal* ones. While external events always need an entity that throws the specific event, internal events may be initiated by the dialogue manager itself. An intelligent doorbell service, for example, might throw an event to switch the focus from an ongoing dialogue to its own spoken dialogue. This more urgent dialogue may tell the user who is at the door. Table 3.3 shows an example dialogue with a system-initiated focus switch. In the example, the interrupting task also influences the user's behaviour: he does not wish to resume the main task. In such a case the ASDM must be able to set an internal event for starting the cooking dialogue later.

However, *external* events may also be thrown by the user. Specific grammars can be used to recognise user commands that are related to a task, which is currently not activated. Table 3.4 shows an example of a user-initiated focus switch. Technically, such focus switches can be handled in a similar way as control dialogues are realised. These are described as part of *Device Adaptation*. Our research focusses on *Event Adaptation* and its two-sided mode of execution in practice. Here, a main challenge is the interruption of dialogues. In the simplest case an external entity would send an event (e.g. an alert message) to the ASDM, which immediately reacts and therefore interrupts any ongoing dialogue so as to prompt the alert message. We assume such behaviour to be uncomfortable and incomprehensible for the user. Thus, we propose to not only react on external triggers but also incorporate internal events. In case an alert message as mentioned above is not life-critical, it should not immediately interrupt an ongoing dialogue. Instead, it may be suppressed until an internal event triggers the system to utter to message. As a trigger the system may, for example, use the time that has elapsed since the external event occurred. If a specific threshold is reached (i.e. we assume the alert gets more urgent during time) there is either no need to suppress the message anymore since the ongoing dialogue has already been terminated or the interruption finally has to be executed. In case of such time-depending actions, for example, the ASDM must be able to react to internal event triggers and to set them accordingly.

There are many reasons for initiating an internal event: an ongoing dialogue must be paused and resumed later, a specific dialogue should only be interrupted if it lasts too long, and the system is able to activate a dialogue that may complement the ongoing dialogue. Technically, several methods for realising such internal event triggers are conceivable: fixed priorities, dynamic priorities (i.e. changing over time), semantic reasoning (i.e. semantically similar dialogues can extend a dialogue), and a individual measure of the progress achieved during an ongoing dialogue. In Sect. 4.2.3.3 we focus on the implementation of fixed and dynamic priorities and in Sect. 5.1 we present the results of their comparison.

The third level of Environment-centred Adaptation is *Task Adaptation* (see Fig. 3.5). During more complex tasks such as the negotiative example discussed in Sect. 3.5, the task itself may vary. Especially during a spoken human–computer negotiation the requirements and/or constraints may change. The effects of *Task Adaptation* may result in task cancellation (which would be a trivial solution) or in slight changes (which may be very complex) such as the extension of an initially planned and predefined dialogue. An example of *Task Adaptation* would be a sub-dialogue that influences an ongoing task driven dialogue. In Sect. 3.5 the main task of the presented scenario is to prepare a dinner. If during a sub-dialogue the system asks the user for specific information that may influence the recipe, the main task has to be modified accordingly. Obviously, *Task Adaptation* corresponds to the most complex level of adaption as it relates to (semi-)automatic learning and/or direct user–system collaboration. In the presented approach we have implemented the fundamental prerequisites of *Task Adaptation* by providing possibilities for disjoint dialogues to influence each other. The main concepts of dialogue variables and inter-dialogue triggers are necessary for this type of intelligent system behaviour. They are explained in Sect. 4.3.2 and in Appendix C.

3.4 Summarising Spoken Dialogue Adaptation

After presenting the underlying ideas of our approach to ASDM, we summarise the seven levels of adaptation in this section. We have recognised the three stakeholders *user*, *IE*, and *SDS* to be the main parties the ASDM has to interfere with. For each of the stakeholders we have defined specific levels of adaptation that require specific ASDM functionalities:

User-centred Adaptation describes two levels of adaptation that concern the emotional state of the user and his specific behaviour.

Emotional Adaptation is not directly part of the ASDM. Here, a main challenge is to characterise and recognise emotions, which is not within the focus of this work. However, in order to be able to properly react on emotions, an ASDM must provide alternative dialogue structures and an open interface that can be accessed, for example, by an emotion recogniser (cf. Sect. 3.6.4).

Behavioural Adaptation is, depending on the definition, a large-scaled field. In this work we focus on the dialogue-related behaviour that can be utilised to adapt the dialogue during runtime. The main aim is to control the spoken dialogue itself. In Sect. 3.5 we present some examples to illustrate the application and take a closer look at *Behavioural Adaptation* in Sect. 3.6.5.

SDS-centred Adaptation comprises two levels of adaptation that define to render the dialogue flow and the spoken interaction itself adaptively.

Dialogue Strategy Adaptation has to be provided by the ASDM so as to optimise the dialogue flow. We have presented strategies to prevent and to repair errors. More details about dialogue strategies are discussed in Sect. 3.6.2 and in Sect. 3.6.2.2.

Speech Adaptation primarily relates to the capability of the ASDM to modify the manner of speech recognition. In case a perfect SDS exists, this kind of adaptation would be obsolete. However, for the improvement of real-life SDSs, methods such as the n-step recognition are of interest. We present technical approaches to *Speech Adaptation* in Sect. 3.6.3.

Environment-centred Adaptation defines three levels of adaptation that concern the changing population of entities within IEs. Those effects that specific events have on a spoken dialogue and the dynamics of tasks are also defined here.

Device Adaptation is provided by the ASDM because of its modular architecture and the implementation of a unified knowledgebase describing the spoken dialogue and its state. Details about this fundamental system capabilities can be found in Sect. 4.3.

Event Adaptation is closely related to *Device Adaptation* but usually leads to a different system behaviour. For instance, under specific conditions, *Event Adaptation* may lead to dialogue situations that are comparable to use cases related to *Dialogue Strategy Adaptation*. In Sect. 5.3 we present the results of a user evaluation concerning strategies related to *Event Adaptation*.

Task Adaptation requires developing an open interface that can be used by the three stakeholders who influence the dialogue. Depending on the actual situation this may lead to intractable system behaviour since the task definition strongly influences the dialogue flow. The underlying concepts of *Task Adaptation* are detailed in Sect. 4.3.2 and in Appendix C.

We have defined this categorisation of adaptation within the context of spoken dialogues for IEs in order to describe the challenges, the requirements, and the functionalities an ASDM should offer. The content of this document is also aligned with the proposed categorisation to allow for a comprehensive and clear analysis of the topic. In the following section we present an application scenario that illustrates the challenges. On the basis of this scenario we furthermore analyse the different levels of adaptation summarised above. In Sect. 3.6 we map the specified levels of adaptation to specific functionalities that are provided by the OwlSpeak ASDM.

3.5 Application Scenario

In this section we present an application scenario that describes and in detail explains the realisation of the ASDM's adaptation capabilities. The user (Suki) is used to digitised spaces and regularly uses his IE, which he calls Julia. She accompanies and observes him and therefore is always aware of his plans. Thus, in our scenario Julia is aware that Suki wants to start the Goal "Prepare dinner for friends" when he arrives at home. To be able to reference the different episodes of the scenario we use the text marks E_1–E_{14} in the remainder of the document.

It is four o'clock in the afternoon and Suki arrives at home. Suki greets Julia (E_1). An emotional recogniser analyses the utterances and sets the emotional state of the user to "happy" (E_2). Julia responds with a personalised greeting (E_3) and then proactively asks him if he wants to start preparing the dinner now (E_4). He agrees and directly enters the kitchen to find out what he can cook for his friends. Due to a fly-over noise Julia does not get the answer at the first attempt and repeats her question (E_5). This time Julia receives Suki's agreement. Julia reminds Suki that one of his friends is vegetarian but eats fish (Suki always forgets about this) (E_6). Suki has no fresh fish available and if possible he wants to avoid going to a grocery store (E_7). A possible solution would be to have pasta with different sauces: Bolognese and a vegetarian one (E_8). Suki agrees "Yes... Switch the lights on!"(E_9). He fills the cooking pot with water, places it on the kitchen stove, and, via speech, switches it to the highest temperature (E_{10}).

Now Suki decides to watch TV since he has to wait for the water to boil. He moves to the living room and asks Julia to switch the TV on. Due to a bad recognition result a correct grammar mapping cannot be achieved, thus the utterance is reanalysed allowing to detect the Keywords "TV" and "on". Because of the uncertain result Julia asks for a confirmation. Suki answers positively (E_{11}). When the water in the cooking pot has almost reached the boiling temperature, Julia decides to inform Suki (E_{12}). Julia now instructs him how to proceed—to add the salt and the pasta to the boiling water (E_{13}). Julia wants to proceed with explaining the recipe but instead she detects that the water with the noodles will overboil if the temperature is not reduced. She explains that she has to switch to a sub-dialogue, the stove

Table 3.5 Exemplary dialogue that may occur as part of our scenario

Speaker	Utterance	Episode	Act
Suki	Hello Julia!	E_1	U_1
Julia	Hi Suki!	E_3	S_1
	Do you want to start preparing the dinner now?	E_4	X_1
Suki	[noise] Yes!	E_4	X_1
Julia	Do you want to start preparing the dinner now?	E_5	X_2
Suki	Yes!	E_5	X_2
Suki	Switch the lights off!	-	U_2
Julia	Simon will attend the dinner. He is vegetarian but he likes fish.	E_6	S_2
Julia	Do you want to go to the grocery before we start preparing the dinner?	E_7	X_3
Suki	No, I really want to start cooking now!	E_7	X_3
Julia	What do you think about pasta with a Bolognese and a vegetarian sauce?	E_8	X_4
Suki	Good idea... Switch the lights on!	E_9	X_4
Suki	Temperature high!	E_{10}	U_3
Julia	Which one? The radiator or the stove?	E_{10}	C_1
Suki	The stove.	E_{10}	C_1
	... [Suki leaves the kitchens and enters the living room.]		
Suki	TV [noise] on!	E_{11}	U_4
Julia	Do you want to turn the TV on?	E_{11}	C_2
Suki	Yes.	E_{11}	C_2
	...		
Julia	The water is boiling.	E_{12}	S_3
	...		

controlling dialogue. Then she asks Suki to reduce the temperature. Afterwards the main dialogue is resumed with an explanation and Julia proceeds with the recipe instructions (E_{14}).

To gain a deeper understanding of the requirements and the knowledge necessary to enable the ASDM to support humans and their collaboration with IEs, Table 3.5 provides an exemplary spoken dialogue that may occur during the presented scenario.

It only represents one possible solution for completion of the task "Prepare dinner for friends" and its related subtasks. The ASDM has to integrate dialogue models that cover a broad variety of alternative dialogues that may possibly occur. Each dialogue act is categorised as user turn (U∗n), system turn (S∗n), or exchange (X∗x). We discuss this categorisation in Sect. 3.5.2. In the following section we extract dialogue act related episodes and map them to the adaptation levels that have been defined in Sects. 3.1–3.3.

3.5.1 Episodes

In this section we focus on the various episodes of this scenario and show how and why the different levels of adaption occur.

E_1 The ASDM must be aware of specific commands and utterance on the part of the user. In this example the system is aware of being greeted by the user. This greeting is defined as part of a specific dialogue that describes various greeting situations. A greeting is important since it may indicate that the user wishes to interact via voice. Since the user therefore controls the spoken dialogue, we categorise E_1 to be part of *Behavioural Adaptation*.

E_2 An external emotional recogniser sets the user state—defined within the user profile—to "happy". Thus, the ASDM drops specific empty phrases that would be useful to motivate the user. In case of a negative emotional user state, motivating phrases such as "I'm so glad, you are back home" can be integrated into all following dialogues until an updated emotional state is recognised. This behaviour requires dialogue descriptions providing alternative dialogue flows that can be chosen during runtime. This behaviour is part of *Emotional Adaptation*.

E_3 A user profile service provides the personal data of the user to the entities populating the IE. In our scenario the greeting utterance may be enriched with the name of the user. This relates to *Device Adaptation*.

E_4 This question–answer pair is proactively initiated by the ASDM. This dialogue turn is defined as part of the IE's task definition defining that it is necessary to proceed with the cooking task. It does not belong to a specific level of adaptation.

E_5 This type of SDS-centred Adaptation is described as *Dialogue Strategy Adaptation*. The system detects an error and changes the dialogue strategy to a so-called "re-prompt" so as to repair the error.

E_6—E_8 These three dialogue steps belong to the complex level of *Task Adaptation*. In the scenario an external planning module [such as PANDA—see Schattenberg et al. (2006) and Bidot et al. (2011)] is necessary in order to align the available ingredients with the requirements of the guests and of the user's preferences. In principle, the ASDM is able to integrate the solutions of such a planner into its dialogue models by using the concept of variables (cf. Sect. 4.3.2).

E_9 Depending on the granularity of the dialogue definition, the dialogue models do not provide grammars that combine an answer to a specific question with a spoken command such as "switch the lights on". Rather, a standard configuration would be to answer the question first and to utter the command subsequently. Such an issue belongs to *Speech Adaptation*. By the use of, for example, keywords to run a second analysis of the user input, this level can be handled by the ASDM.

E_{10} This behaviour is part of *Device Adaptation*. The IE detects the necessity to activate the stove. As a result it triggers the ASDM to add the dialogue model of the stove to its knowledgebase and to integrate it into the dialogue generation.

E_{11} This example shows an aspect of *Speech Adaptation* in combination with *Dialogue Strategy Adaptation*. After the recognition fails, the ASDM is able to understand parts of the commands by using keywords. Under these circumstances, this kind of n-step understanding is fraught with uncertainty. Hence, the ASDM initiates a confirmation dialogue and therefore adapts the dialogue strategy.

E_{12} Depending on an external trigger the ASDM alerts the user of an urgent event. This behaviour is described by *Event Adaptation* and is technically realised by switching the dialogue focus to the urgent dialogue.

E_{13} This dialogue steps are part of a prescribed dialogue model that is used to explain the user how to prepare the dinner.

E_{14} The final part of the scenario shows a combination of *Event Adaptation* and *Dialogue Strategy Adaptation*. Depending on an urgent external event the ASDM has to pause the ongoing dialogue. The ASDM switches the focus to a short dialogue so as to prevent the water from overboiling. Such topic or focus switches are not trivially to be realised. In Sect. 3.6.2.1 we present several strategies to provide an effective and at the same time user-friendly topic switch.

The exemplary episodes are meant to be used as a guideline of what functionalities an ASDM must provide. In the remainder of this document we refer to the episodes so as to illustrate a practical example of the specific functionality.

3.5.2 Conversational Acts

The scenario presented in the previous sections seems to be straightforward. However, after taking a closer look at the plot, the various levels of adaptation as defined in the Sects. 3.1–3.3 become visible. To develop a general strategy for generating adaptive dialogues, we present a classification of conversational acts that are minimally necessary to perform the conversation that is related to the scenario. In the remainder we use the terms *conversational act* and *dialogue turn* interchangeably. In order to allow for both a complete and efficient modelling of spoken dialogues it seems to be expedient to define a set of generic building blocks. These can be used to generate the conversation. Table 3.5 distinguishes the conversation between the user (Suki) and the system (Julia) into three types of fundamental conversational acts: user turns (U∗n), system turns (S∗n), and exchanges (X∗x). Analogously to the definition in ITU (2005) we define them as follows:

User turn (U) The first utterance in Table 3.5 is a user turn (U_1). The system waits for a specific input from the user before the conversation starts. In our example the system is explicitly not allowed to proactively start the conversation. In the most cases, spoken commands such as $U_{2,3,4}$ are also realised as user turns.

System turn (S) Utterances such as greetings, warnings, or single statements are generated as system turns ($S_{1,2,3}$). This relates to the idea of sharing knowledge between system and user as described in Larsson (2002). In our scenario we do not use explicit confirmations on part of the user but assume that he correctly receives the information the system emits.

Exchange (X) An exchange relates to questions or statements by one interlocutor that requires an answer by an opponent interlocutor. There are two cases of exchanges: a single statement on the part of the user and a system-directed answer. The former requires an answer from the system and the latter refers to a question the user answers. The utterances and the grammars are encapsulated as *moves*. We define this fundamental concept of the OwlSpeak ASDM in Sect. 4.3. The first case would be generated as a user turn combined with an additional system turn. The latter case would be generated as a specific set of moves, whereby a single move provides an utterance that is utilised as the system's question and a set of grammar moves that might match a possible answer by the user ($X_{1,2,3,4}$).

A further type of conversational acts are so-called *clarifications* ($C{*}n$). The example shows two types of clarifications as an example: C_1 and C_2. Whenever the ASDM is not certain about a specific user input, the ASDM automatically generates a specific clarification dialogue. In general these dialogues are realised as exchanges. However, they are not explicitly defined as part of the dialogue model and are therefore introduced as a separate class of conversational acts. We distinguish two major types of dialogues the ASDM generates so as to clarify a user input:

Decision-making turn Turn C_1 is generated since the user input matches two grammars for different devices: the radiator and the stove. The user does not provide sufficient information to allow for automatic decision on which dialogue the utterance refers to, i.e. which device or service the user wishes to control. Hence, the system must initiate an exchange asking the user for clarification. In case of two to five similar entities understanding similar commands it is feasible to list the "friendly names" and to ask the user to choose which one to be controlled. However, if the number of alternatives exceeds seven, we assume the dialogue would confuse the user (cf. Sect. 4.2.3.2).

Confirmation turn Turn C_2 is automatically added to the conversation since the previous command could not be matched by the grammar. A second attempt using keywords (cf. Sect. 3.6.3) is necessary. Since such an analysis is always more fraught with uncertainty than a direct grammar match would be, the system has to assure its detection. Results of our work implied that it is beneficial to add a confirmation turn instead of ignoring the user's input or carrying out an uncertain command. Exactly these uncertain commands lead to misunderstandings, which are disruptive to human–computer spoken interaction.

In Sect. 4.2.3.2 we present technical details regarding these clarification dialogues. For the proposed approach and the ontology-based modelling of spoken dialogues the introduced taxonomy seems to be crucial since the different conversational acts result in different conversational strategies. In our approach, these strategies result in selecting different dialogue templates in order to generate the subsequent dialogue steps (see Sect. 4.3.3). After having analysed the dialogue structure, in the following section we explain the characteristics of the ASDM's functionality and the corresponding features that are necessary to provide the proposed seven levels of adaptation.

3.6 Characterisation of the Functionalities

The presented functionalities are required by the seven levels of adaptation and can directly be mapped to one or more levels:

- Capability to provide multi-tasking spoken dialogues (cf. Sect. 3.6.1)

  → *Behavioural Adaptation*
  → *Dialogue Strategy Adaptation*
  → *Device Adaptation*
  → *Event Adaptation*
  → *Task Adaptation*

- Capability to provide specific dialogue strategies (cf. Sect. 3.6.2)

  → *Dialogue Strategy Adaptation*

- Capability to influence, i.e. to adapt the recognition behaviour of the SDS (cf. Sect. 3.6.3)

  → *Speech Adaptation*

- Capability to provide speaker aware dialogues (cf. Sect. 3.6.4)

  → *Emotional Adaptation*

- Capability to allow the user to control the dialogue (cf. Sect. 3.6.5)

  → *Behavioural Adaptation*

- Integrability into an IE, e.g. into the ATRACO System (cf. Sect. 4.1 as part of the next chapter)

  → *Device Adaptation*
  → *Event Adaptation*
  → *Task Adaptation*

In the following sections we focus on the functionalities and refer to the episodes we have illustrated as part of the application scenario. The functionalities of the ASDM are then summarised in Sect. 3.7.

3.6.1 Multitasking Spoken Dialogues

One of the most important features of the ASDM is its capability to manage more than one dialogue task in parallel. State-of-the-art SDSs for telephony-based or in-car applications are designed to perform a single task or a predefined set of tasks. However, similar systems for IEs have to deal with a more complex situation. One of the main sources of complexity is the multi-domain and thus multi-topic nature of real-life processes. If the dialogue domain is not clearly defined, it seems almost impossible to collect an appropriate corpus or to define valid rules to control the dialogue flow. In our approach we propose to use independent and/or interdependent dialogue models to encode both the dialogue structure and its state. This allows easily activating, deactivating, pausing, interrupting, and resuming of tasks. Depending on the activated set of dialogue models, the spoken interface provides different spoken dialogues. The idea of utilising multiple dialogue models to provide an adaptive spoken dialogue interface can be compared with the concept of "widgets". These are used within the context of GUIs (Cáceres 2011). Hence, depending on the actual necessity, the user or the IE may activate or deactivate specific spoken dialogue widgets so as to "mash-up" the interface. Multitasking spoken dialogues can be seen as a high-ranking functionality that cannot be directly related to a single level of adaptation. It rather relates to most of the defined levels and more specifically to *Behavioural Adaptation*, *Dialogue Strategy Adaptation*, and all environment-centred levels of adaptation. In our scenario, Episode E_{10} is an example of a functionality that requires multitasking (see Sect. 3.5). The IE detects the necessity to active the stove and thus triggers the ASDM to add the dialogue model of the stove to its knowledgebase and to activate it.

If the dialogue management should not solely be performed by the user, a main question that arises is *how to automatically decide which dialogues to activate or to deactivate*? In the example of the ATRACO IE, the Interaction Agent (IA—see Sect. 2.2) receives a trigger from the sphere manager to initiate or to terminate a dialogue. The IA decides whether a graphical or a spoken dialogue or both interfaces can be instantiated. Of course, various other options to activate and to deactivate a dialogue are feasible. External triggers such as timers, motion sensors, or events caused by third persons may influence the spoken interface. If more than one spoken dialogue can be provided by a speech interface (i.e. one set of speakers and microphone) in parallel, a proper handling of these *concurrent* dialogues (see Sect. 4.2.3.1) is challenging. It becomes necessary to define rules or insert priorities to select the appropriate grammars and the most important utterance at a specific point in time. Obviously, utterances can only be provided one after another whereas grammars can be activated in parallel. However, overlapping or similar grammars have to be detected beforehand in order to avoid misunderstandings. If this is not possible, proper strategies have to be developed to repair the misunderstanding, i.e. the *conflicting* dialogue (see Sect. 4.2.3.2). Within the context of IEs we have recognised four multitasking situations that are treated differently:

Various command-and-control dialogues These dialogues typically consist of several grammars. Usually, each grammar defines a specific set of commands. Within an IE, commands for controlling a light, the heating, and several other devices and services may be controllable by these dialogues. The grammars provided by the devices and services can be interpreted in combination with each other. In case of ambiguities a conflict solving dialogue has to be initiated (cf. "conflicting dialogues" in Sect. 4.2.3.2).

One main dialogue combined with command-and-control dialogues Only the grammars that are needed to control the IE are added to the main dialogue. This dialogue typically consists of adjacency pairs that can successively be processed. In case of ambiguities the main dialogue suppresses the command-and-control dialogues automatically (see "concurrent dialogues" in Sect. 4.2.3.1).

One main dialogue combined with sub-dialogues The sub-dialogues must be proactively initiated by the user or automatically by the IE. If the IE initiates such a dialogue, the ASDM may judge its urgency and either suppresses the ongoing main dialogue or stalls the sub-dialogue until the main dialogue has been terminated. A dynamic approach towards combining these two ways of dialogue combination is presented in Sect. 4.2.3.3.

Various main dialogues In order to avoid confusing the user, the dialogue models are usually designed to prevent the dialogue logic to combine more than one main dialogue. By utilising priorities, a main dialogue, for example, for booking a hotel, would be terminated before a second main dialogue, for example, for booking a flight would be initiated. However, the OwlSpeak ASDM framework provides functionalities to influence a follow-up dialogue in case the first dialogue has collected relevant information (see Appendix C).

Depending on the type of dialogue situation, the ASDM derives from the dialogue models, the dialogue logic decides which multitasking strategy to be applied. However, besides developing the technical way the systems decides how to provide multiple spoken dialogue tasks, it is also necessary to assist the user in case the dialogue switches its focus. Thus, in the following section we provide information about the dialogue strategies that have been developed as part of the ASDM framework.

3.6.2 Dialogue Strategies

In this section we present two types of dialogue strategies that are important for *Dialogue Strategy Adaptation* as part of SDS-centred adaptation. These strategies have been integrated into the prototype, technically tested, and evaluated under realistic conditions. In the following section we present the ASDM functionality and provide examples that are related to the application scenario.

3.6.2.1 Topic Switching Strategies

The proposed ASDM solves the issue of switching between several different dialogues or tasks. However, to deliver technical solutions to a specific problem and to behave in a human-like manner are two different things. Especially when it comes

Table 3.6 An exemplary dialogue applying negotiated task control to the short reminder

Speaker	Utterance
Julia	Do you want to go to the grocery before we start preparing the dinner?
Suki	No, I really want to start cooking now!
Julia	*Sorry, may I interrupt the dialogue for a moment?*
Suki	Yes.
Julia	The washing machine is done.
Suki	OK.
Julia	What do you think about pasta with a Bolognese and a vegetarian sauce?

to switching between independent tasks or topics during an ongoing conversation, no adequate solutions seem to be available so far. The way interruptions are handled influences the cognitive resources and the information management capacity of humans. So as to assist the user during a multitasking situation, we present different methods of how dialogue interruptions can be handled. These topic switching methods can automatically be applied by the ASDM to the dialogue depending on the actual situation, the urgency of the dialogue turns, and the user characteristics. This kind of dialogue enhancement is related to *Dialogue Strategy Adaptation* (cf. Episode E_{14} in Sect. 3.5). We have conceptualised four different strategies to switch between tasks based on the previous work presented in Sect. 2.5.2 and on our own hypotheses. The starting point of our investigation was a main task and several subtasks the ASDM switches to during the ongoing conversation. The default system behaviour is to immediately switch to a more urgent topic without any assistance. We used this as a baseline since it seems to be the most common way of task switching: an unassisted immediate topic shift. The system acts regardless of any cognitive limitations of the user. Neither warning signals for switching between tasks nor context recovery after the interruption has been provided. In the following we present the strategies that have been applied to the prototype. Three interrupting tasks cross over the main task one after another. During the main task the user and the system discuss the preparation of a dinner (cf. Sect. 3.5). In between, a short notification as presented in Table 3.6 interrupts the dialogue.

Furthermore, a long reminder with detailed additional information (see Table 3.7) and a complex subtask (see Table 3.8) interrupt the main dialogue. The objective of this conversational situation is to investigate the correlation between the user's ability of maintaining situational awareness and a preferred form of task interruption. Three out of four strategies immediately integrated the interruption into the main dialogue. Here, our approach focusses on mechanisms providing efficient task switches such as attention markers and task resuming after the interruptions occurred.

To introduce a more sophisticated strategy, we use discourse markers combined with prosody cues for signalling the switching of tasks. With conversational cues such as "wait", "oh", and "well" we intent to catch the user's attention for the upcoming interruption. A higher pitch and volume at the beginning of the interrupting task are ought to extent the situational awareness to more than one

Table 3.7 An exemplary dialogue applying discourse markers (in italics) to a detailed long reminder

Speaker	Utterance
Julia	Do you want to go to the grocery before we start preparing the dinner?
Suki	Yes.
Julia	*Ah!* Your friend Mario celebrates his birthday on Thursday. You should buy him a present.
Julia	*So,* what do you think about pasta with a Bolognese and a vegetarian sauce?

Table 3.8 An exemplary dialogue applying the *explanation* strategy to the subtask

Speaker	Utterance
Julia	Do you want to go to the grocery before we start preparing the dinner?
Suki	Yes.
Julia	*Let's talk about something different.* You are supposed to meet Sandra for dinner on the 10th of March. Do you want to make a reservation now?
Suki	Yes.
Julia	You can choose between "San Marco", "Enchilada", and "Chinese World".
Suki	I prefer "Enchilada".
	...
Julia	*Well, let's continue with preparing the dinner.* What do you think about pasta with a Bolognese and a vegetarian sauce?
Suki	Good idea.

task. The discourse markers "so" and "further" are used to assist the user to resume the main task. An exemplary dialogue applying this strategy is shown in Table 3.7. A further strategy we propose is to employ full sentences initialising topic shifts instead of short discourse cues. These sentences seem to produce a more natural dialogue flow and increase the user's time span for task switching detection and realisation. Complex task recovery to the primary task that has been interrupted is provided by the system as well. Table 3.8 demonstrates the implementation of this enhanced way of task switching.

The fourth strategy we have investigated supports negotiated task interruption in the form of system queries such as "May I interrupt you?". Here, the system allows the user to have full control of the task delivery. In case of a positive confirmation, the system switches to the interrupting task. If the user denies, the main dialogue is continued. Table 3.6 shows the mechanism of the negotiated task switching control. The three described strategies seem to give a competitive edge compared to the baseline system behaviour regarding the human cognitive resources they require. However, they may also have some drawbacks. The precise explanation of each topic shift may be boring and even frustrating to the user. In some cases, the ambiguities of the discourse markers "so" and "than" may lead to confusion. By negotiation some critically important tasks may be rejected. In order to prove the efficiency and the reliability of the presented strategies we have conducted a study to test the approaches during a real user–system interaction with the proposed ASDM (cf. Sect. 5.3). In the next section we present repair strategies that are also part of *Dialogue Strategy Adaptation*.

3.6.2.2 Repair Strategies

In Sect. 2.5.2 we have discussed former research on repair strategies. The authors we mentioned mainly focussed on the efficiency of the different strategies they investigated. However, the question of usability and user-friendliness is equally important. Bohus and Rudnicky (2005) recognised the most effective strategies regarding dialogue success rate. Although this is a crucial point we argue user satisfaction to be of similar importance. If a system annoys, confuses, or bores the user because of an inappropriate strategy, the dialogue flow is disturbed. Hence, the user-friendliness is lowered even if the success rate remains the same. We propose to consider the expected user satisfaction when choosing a repair strategy. For the experiments presented in Sect. 5.4 we focus on the following three strategies:

Modified MoveOn The strategy (Skantze 2003) proposed seems to be best suited for route guidance and similar applications. However, without modifications it cannot be adapted to SDSs for collecting information as exemplified in the application scenario. In case the information to be collected is interrelated, the MoveOn concept can be applied. It may be more efficient to skip explicit confirmations for several consecutive questions and to ask for a combined confirmation. In case the user rejects, it is possible to detect a misunderstanding using on a combined score.
This score may include the n-best lists of the ASR, the dialogue history, and general reasoning to exclude unlikely combinations. An example of MoveOn would be a system asking the user three questions without repeating the input of the user. Instead the system initiates an explicit confirmation summarising the answers to all three questions. If the user does not confirm, the system is able to "guess" a correct combination by repairing user input of low confidence. An example of a spoken dialogue utilising this strategy is presented in Table 3.9.

Re-prompt After each dialogue step, the utterance is immediately implicitly confirmed. In case of a misunderstanding the user must explicitly interfere with signal words such as "wrong" or "false". This triggers the system to inform the user of the desired correction, followed by a simple re-prompt of the original question (cf. Episode E_5 in Sect. 3.5).

Help This strategy works similarly as the re-prompt. However, a help dialogue providing all information required to complete the specific dialogue step is added.

Table 3.9 An example of a successfully completed MoveOn Dialogue with a system failure

Speaker	Utterance
Julia	What pasta sauce do you want?
Suki	Bolognese sauce. ["Bolognese" understood]
Julia	What ingredients do you want for the vegetarian sauce?
Suki	Tomatoes. ["courgettes" understood]
Julia	What else should be added or say "done" to finish the selection.
Suki	Aubergine. ["aubergine" understood]
Julia	What else should be added or say "done" to finish the selection.
Suki	Done.
Julia	I have Bolognese and a vegetarian sauce with courgettes and aubergine. Is that correct?
Suki	No.
Julia	Oh, I have one last guess, else we have to start over. I have Bolognese, a vegetarian sauce with tomatoes and aubergine! Is that correct?
Suki	Yes, that's correct.
Julia	Great, so let's start cooking.

In the following section we investigate technical solutions to optimise the spoken user–system communication. We adapt the recognition process of the SDS in order to allow the ASDM to properly react on unwanted system behaviour.

3.6.3 Adaptive Understanding

As discussed in Sect. 2.5.1 non-understandings may significantly reduce the user-friendliness of SDSs. Within the framework of the ATRACO Project a qualitative social evaluation series has been conducted within the iSpace[1] at the University of Essex (cf. Sect. 5.5). This study revealed a main lack of our model-based approach to SDM: even the most sophisticated model cannot cover all contingencies. As a reaction to the performance the system showed in the field we investigate the possibilities to increase the recognition rate of the model-based ASDM. As described in Sect. 4.4 the ASDM uses on-the-fly generated VoiceXML documents to describe a currently valid dialogue. These dialogues are newly generated every 3–10 s, depending on the set-up of the ASDM.

Since it is impossible to manipulate the grammar while the interpreter parses the VoiceXML document, we propose to perform a second analysis of the user input in

[1] http://ieg.essex.ac.uk/?page_id=32

Table 3.10 Exemplary dialogue showing the method of adaptive understanding

Speaker	Utterance	Grammar-based recognition	Keyword-based analysis
Suki	TV [noise] on!	[→ grammar fails]	
		–	[→ keywords match "TV"]
		–	[→ keywords match "on"]
Julia	Do you want to *turn the TV on*?	–	[Confirmation turn]
Suki	Yes.	[→ grammar match]	

case the grammar fails, i.e. in case of a non-understanding. To be more specific: we use a nested keyword-spotter to analyse the user input after the grammar has failed (Rohlicek et al. 1989). The ASDM uses a set of dialogue domains to represent the various tasks and topics within an IE. One option is to design one dialogue model per domain (i.e. per device or service). These domain models provide an additional set of keywords that can be used to detect the actual domain. Thus, a model that describes the possible (spoken) commands that can be used to control, for example, a lamp as part of an IE, may provide keywords such as "light", "lamp", "shiner", and "luminary". These keywords are used by the nested keyword-spotter to find out to which domain the utterance of the user may belong to. An example of the described technique is presented in Episode E_{11} in Sect. 3.5. We have detailed the crucial parts of this episode in Table 3.10.

The presented dialogue situation starts with a noisy command by the user. The ASDM receives an event signalising a non-understanding combined with a record of the user's utterance. The system initialises a second analysis of the input by utilising the keywords that are defined as part of the dialogue model. The system detects to which model the utterance possibly refers to: to the TV domain. A second analysis can then be started by utilising keywords that are specific for a command within the detected TV domain. Such keywords might be "on", "off", "next channel", or "volume up". If this analysis concludes with a valid result, i.e. "domain=TV" and "command=on", the ASDM generates a confirmation dialogue and asks the user if he wishes the TV to be turned on. Valid answers are "yes" or "no". On one hand such a procedure avoids the necessity of a second user input that may result in a further non-understanding. On the other hand the system does not end up performing the wrong command automatically because of a misleading combination of keywords.

An alternative approach that does not use a fixed (limited) grammar but an extensive dictionary is referred to as *out of vocabulary* recognition (OOV). As a result the system would receive, for example, a n-best list of results from the recogniser. Obviously, due to the lack of a domain grammar, this may easily lead to confusion. There is, for example, a high probability that two similar-sounding words such as "house" and "mouse" may be mistaken. Thus, we propose to use the complete n-best list of freely recognised utterances to analyse the user intention. In other words, the main task is to determine which semantic meaning is connected to the utterance. This could be performed by a string-based comparison of the n-best list and the words listed as keywords (if the grammar-based recognition has not resulted in an understanding before).

This, however, would lead to a significant number of mistakenly detected non-understandings. For example, if the word "light" is listed as part of the keywords but the user's utterance could not be detected in the first attempt, an OOV recognition would provide a n-best result listing {might, flight, right,...}. The ASDM would still (after the second analysis) only be able to emit a non-understanding. Especially homophones are problematic within this context. To encounter this, we propose to *blur* the keywords by adopting the Levenshtein distance (Levenshtein 1966). This distance is described as the minimal number of insertions, deletions, and substitutions of characters needed to transform one string into another. In order to detect homophones that have erroneously been recognised, the Levenshtein distance of the recognised word and the words that are part of the grammar can be calculated pair-wise. For the German language a distance of one would be sufficient here. For the English language there are several homophones with a higher distance (e.g. "colonel" and "kernel"), which would admittedly lead to more confusions. Thus, we propose to adopt a low distance so as to benefit from the *blurred* keywords without provoking too much misunderstandings.

A further attempt that utilises OOV recognition is the semantic grammar interpretation. A dialogue model for handling a greeting situation may provide "hello", "good morning", and other salutations. However, if a specific salutation such as "hi" is not covered by the grammar (note that we utilise a minimal grammar in order to reduce overlapping commands and misunderstandings), the system cannot react appropriately. Thus, we propose to figure out the semantic meaning of "hi", which can then be mapped on the semantic value "salutation". This is encoded as part of the dialogue model to be the semantic meaning of "hello". As external semantic knowledgebase we utilise GermaNet (Hamp and Feldweg 1997), a lexical-semantic dictionary. GermaNet is the German version of WordNet (Miller 1995). It provides relations to detect synonyms, hyponyms, meronyms, and hypernyms. In Sect. 4.2.3 we present the technical implementation of the advanced understanding methodologies that have been introduced hitherto.

The strategies and methodologies presented in this and the previous sections are useful. They improve the communication between user and system with respect to a suboptimal SDS, which is a realistic assumption. As mentioned, in case of an optimal SDS, the ASDM would not have to offer advanced understanding methods or enhanced dialogue strategies. Nevertheless, we may also apply this idea to the user–system communication with respect to a not *suboptimal user*: in the following two sections we provide information about speaker aware dialogues and about a method that allows the user to take control of the spoken dialogues that are provided by the SDS.

3.6.4 Speaker Aware Dialogues

As described in Sect. 3.1, one aspect of User-centred Adaptation is *Emotional Adaptation*. Emotion recognition is an application of speaker awareness. In Schmitt et al. (2009) we have elaborated speaker aware VoiceXML Dialogues. Such a dialogue is able to recognise, for example, the emotional state, the age, and the gender of a user. As explained in Sect. 4.4 the proposed ASDM generates VoiceXML dialogue snippets that are compatible to our approach to speaker awareness. Furthermore, the dialogue models the ASDM utilises are perfectly suited to incorporate alternative dialogue flows that can be selected depending on the current speaker state. User studies, backing up the hypothesis that dialogue systems tailored to specific user groups are considered to be more user-friendly, can be found in Metze et al. (2008) and Burkhardt et al. (2008). Potential applications for user group aware dialogue systems are numerous. According to the taxonomy of Burkhardt et al. (2007), for this technique different application scenarios are conceivable:

Direct address. Especially elderly people tend to address an SDS with "Sir" or "Ma'am". The system may behave similarly and address users directly, e.g. "This is obviously my fault, Sir!" and "One Moment, Ma'am, I'm checking the available recipes".

Dialogue strategy adaptation. Dialogue prompts, dialogue strategies, and dialogue grammars may be adapted to the respective user group and the emotional state. Elderly people may experience more direct confirmation prompts, teens could be addressed informally, and to angry users the system may react with a conciliation strategy.

Problematic dialogue prediction. Task completion rate prediction as described in Herm et al. (2008) and Schmitt and Liscombe (2008) can detect critical situations in the dialogue and may trigger the ASDM to intervene before the user terminates a dialogue without completing the task. By taking a priori information about the user group into account, this prediction can be more exact, provided that there are differences between user groups in completing the task (e.g. expert vs. novice users).

Episode E_2 in Sect. 3.5 illustrates an example of the effect an emotional recogniser may have regarding an ASDM-controlled SDS. However, to establish an emphatic SDS several challenges remain. Currently, we created and tested the user state models with "artificial" corpora containing clean speech. Here, reliable annotations for age, gender, and emotion already exist, which is an advantage. However, performance losses are to be expected when this approach is used in real-life applications. First, the corpora consist of acted emotions, which are not comparable to spontaneous emotions. Second, the utilised corpora do not contain noise, which is atypical for real-life data. In the following section we focus on the second level of user-centred adaptation: *Behavioural Adaptation.* While the important mechanisms of *Emotional Adaptation* are almost completely located outside the ASDM, *Behavioural Adaptation* relates to inherent dialogue management tasks.

3.6.5 Voice-based Dialogue Control

Nowadays, SDSs are usually developed as task-specific interfaces. Compared to a windows-based GUI an SDS provides the functionalities of the program itself and

Table 3.11 The user activates a specific dialogue by using a *system command*

Speaker	Utterance	Action
Suki	Lights on!	[no action]
Suki	*Activate lights control!*	[light controlling dialogues activated]
Suki	Lights on!	[Julia switches the lights on]

not the tasks of the operating system (OS). Possible tasks are to close programs, to open programs, to switch to other programs, and to minimise programs. Since there are several parallel tasks within the IE context, it seems to be crucial for the ASDM to provide mechanisms to allow the user to control the spoken dialogues in a similar manner as an OS does. An example of voice-based dialogue control is illustrated in Table 3.11. The user activates a specific spoken dialogue by triggering the ASDM to incorporate the dialogue model for controlling the lights. In Sect. 4.2.1 we provide an overview on all commands that can be used to control the ASDM. It is possible, for example, to activated or to deactivated dialogues (see Table 4.2). As the application scenario illustrates, a greeting may also be used to start a specific dialogue. Technically, the ASDM is triggered by the greeting to activate a specific dialogue model and to add it to his dialogue knowledgebase (cf. Episode E_1 in Sect. 3.5).

However, many other applications of voice-based dialogue control are also conceivable. According to Bohlin et al. (1999) a very important dialogue control task for the user is to repeat a system utterance on request and to ask for a reformulated output. In principle, the ASDM is able to provide this kind of user-controlled spoken dialogue. Of course, the dialogue designer must define specific grammars for the various dialogue control related commands. If the user is aware of the "friendly names" of the dialogue domains the ASDM covers, he is able to utter "Activate *light control*" in order to activate the spoken dialogue that offers lighting control. In the remainder we distinguish between commands that are domain related, i.e. that offer the functionality the dialogue has been modelled for and so-called *system commands* that can be used to control the overall ASDM behaviour thus providing *Behavioural Adaptation*. In the following section we conclude this chapter and discuss the characteristics of the presented functionalities.

3.7 Conclusion

Before we dive into the technical details of the OwlSpeak ASDM in the following chapter, we summarise the definitions and classifications that have been established so far. Figure 3.7 shows an overview on the relations between the ASDM, the levels of adaptation, and the concrete functionality. The proposed prototype, the OwlSpeak ASDM, connects the three stakeholders; the user, the SDS, and the IE.

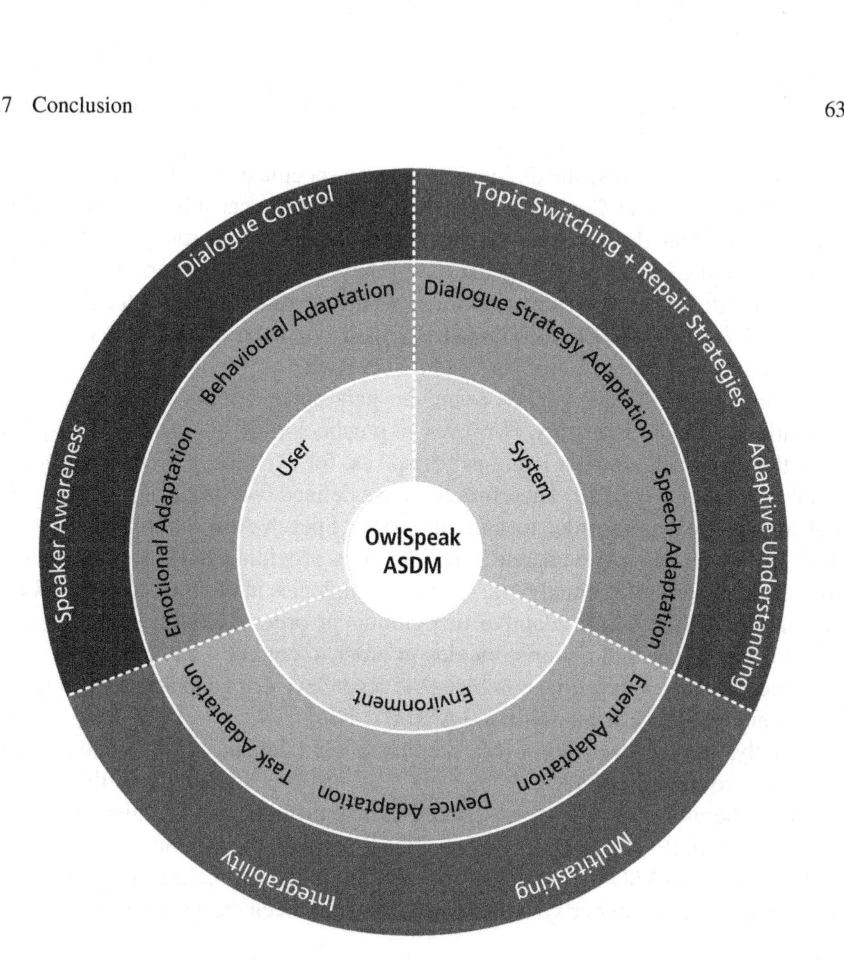

Fig. 3.7 A functional circle showing the seven levels of adaptation categorised by the three stakeholders and mapped to the specific functionalities

We argue that the environment in this case refers to the application as categorised in classic SDS terminology. The main difference between the *environment* and such an *application* is that the features and the resources of the environment may change, which is not the case for an application. We have classified three levels of adaptation describing the relation between the ASDM and the environment: Event, Device, and Task Adaptation. These adaptation levels allow to reveal the functionalities that are necessary to provide coherent dialogues within a changing domain. Two main challenges arise from this special nature of the IE. For one thing the ASDM must be integrable in general, and for another thing it has to be capable of multitasking. The former implies to cope with the information exchange that happens within the environment and the latter requires to switch between more than one task and, in order to be able to do this, to save the dialogue states persistently.

Of course, the most important stakeholder is the user himself. We have revealed two levels of adaptation to define the relation between the user and the ASDM: Behavioural and Emotional Adaptation. The former relates to the user's ability to control the dialogue flow: we propose to use a limited set of meta-commands that can be used to, e.g., pause the dialogue or select a specific dialogue domain. We call this feature "Dialogue Control". The latter refers to the capability to incorporate information about the speaker: his age, mood, and his expertise are of interest from the point of the ASDM. Speaker awareness is of upmost importance in this context. We do not focus on how to extract this information. However, the developed framework is able to handle such knowledge if an external module is able to provide it. The third party we introduce here is the SDS itself.

As mentioned, if the ASDM would be part of an optimal SDS, no related adaptation would be necessary. However, in practice, SDSs are usually error-prone since the spoken interaction is not as "crisp" as, for example, a GUI or keyboard-based interaction would be. Therefore, we have defined two levels of adaptation the ASDM must realise in order to face this issue: Speech Adaptation and Dialogue Strategy Adaptation. As concrete functionalities providing these levels we have investigated and implemented topic switching strategies in order to prevent failures and unwanted behaviour, adaptive recognition in order to cope with erroneous dialogue situations, and repair strategies in order to correct a flawed dialogue. By implementing a prototype that combines the seven levels of adaptation, we describe a novel attempt towards model-based ASDM.

Notably, our approach is not restricted to specific domains or dialogue applications but provides a complete framework that supports adaptation in the defined manner. Therefore, it is perfectly suited for controlling and defining a spoken interface within IEs. In the following chapter we explain the technical fundamentals of the OwlSpeak ASDM framework, present the technical realisation, and further detail the main ideas underlying the system that have been discussed in this chapter.

Chapter 4
The OwlSpeak Adaptive Spoken Dialogue Manager

The conceptual basis of the OwlSpeak ASDM is explained in this chapter. We discuss the prototype and describe the implementation of the functionalities that have been derived from the seven levels of adaptation presented in Chap. 3. The functional principles are also explained in detail. Therefore, we focus on the technical realisation of adaptiveness in relation to spoken dialogue within IEs. We present the ontology-based framework for defining and modelling spoken dialogues and show how to integrate both *structure* and *state* of spoken dialogues into a unified model.

4.1 Architectural Overview

In Chap. 3 we have identified adaptability to be the main requirement for an ASDM within IEs that are in turn characterised by changing domains. In order to meet this requirement, OwlSpeak is implemented considering the Passive View variation of the Model-View-Presenter (MVP) pattern (Potel 1996; Fowler 2006). MVP is a model-derivative of the popular Model-View-Controller (MVC) paradigm (Reenskaug 1979; Krasner and Pope 1998). The underlying idea of MVC is to divide an application or a system into three logical parts as illustrated in Fig. 4.1.

The user links between view and controller by raising events and by perceiving the data provided by the view. The model, as defined by the MVC pattern, provides the basic logic required to update the view. The main functionality of the controller is to manipulate the model in order to react on user input. In classic MVC, the view queries the model, thus providing basic logic as well. However, for the OwlSpeak ASDM we preferred a more rigid distinction between view and model since we wanted to foster the testability and the integratability of the system. Therefore, we implemented OwlSpeak using the MVP Passive View derivative of MVC. As shown in Fig. 4.2 the user solely interacts with the view layer. Contrary to MVC, the presenter mediates between model and view—the model conveys no functionality,

T. Heinroth and W. Minker, *Introducing Spoken Dialogue Systems into Intelligent Environments*, DOI 10.1007/978-1-4614-5383-3_4,

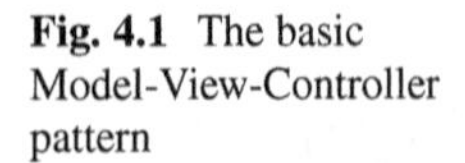
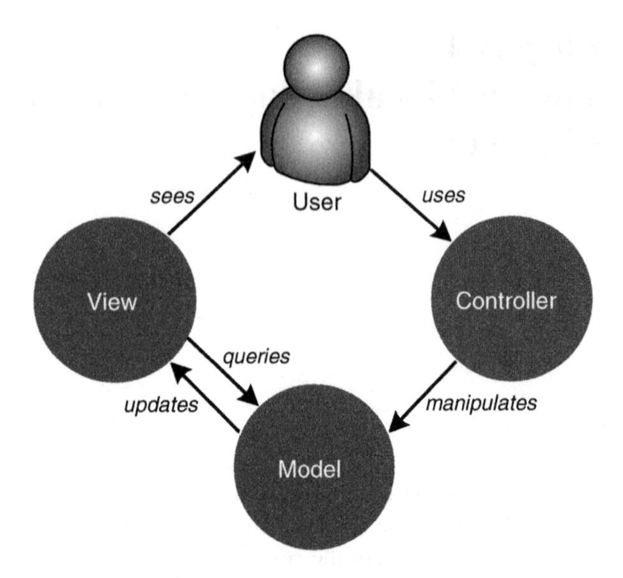

Fig. 4.1 The basic Model-View-Controller pattern

i.e. it is not an application but encodes the knowledge that is used by the presenter. The term *model* in this case refers to a *domain model*. The view behaves passively and allows for a flexible system that can be tested easily on a per-module basis.

Therefore, especially for model-driven systems that directly interface to the user MVP is perfectly suitable. To allow the communication with a user or with other external entities the application needs a knowledgebase that describes facts and the relation between such facts. A fact, for example, is the name of a person or an ID number. A relation such as "has" can be used to express:

person 'A' *has* ID_number '4711'.

Without this knowledge the system is unable to generate useful output, i.e. act as *knowledge source*. Furthermore, it would not be able to understand input that is provided by external entities, thus acting as *knowledge sink*. We define *to understand* in this context as *to bring into relation* with existing knowledge. The term *domain model* is specified as the knowledge a system needs for the interaction with the user within a specific context. There are many ways to establish such a knowledgebase: to name but a few, SQL databases or XML files can be utilised. A more sophisticated option that we use for the prototype is to take advantage of OWL ontologies for the provision of a common understandable knowledgebase. In practice, the user does not directly interfere with this machine-readable model. In fact, the MVP pattern describes the view as the only layer that interacts with the users. To render the information into a human-perceivable form, for example, a table, a diagram or, in our case, a specific spoken dialogue description is passed to a frontend such as a screen (for the provision of a GUI) or an SDS (for the provision of a Voice User Interface (VUI)). A main benefit of separating the view from the other layers is that the application may provide more than one input and output layer (i.e. view), which can even be changed or modified during runtime. Thus,

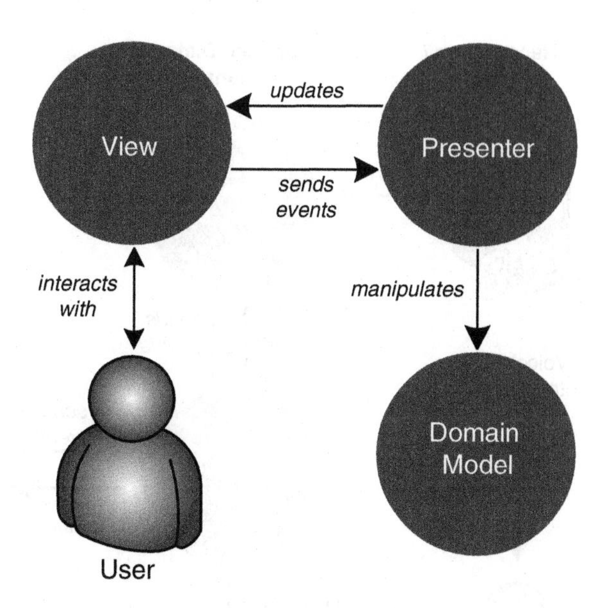

Fig. 4.2 The MVP Passive View pattern as used in OwlSpeak

the ASDM is independent of the dialogue description. The prototype generates VoiceXML; however, other description languages or proprietary formats are also feasible (cf. Sect. 4.4).

Obviously, the third layer described by the MVP paradigm, the presenter, acts as link between the model and the view. On one hand, it has to translate the knowledge described by the model into a format that can be utilised by the view to generate meaningful (user-)interfaces. On the other hand, it has to react to input that is initiated by the user. This input must be translated into a format that corresponds to the underlying knowledgebase, the model. According to changes in the model, the view is also updated by the presenter. Hence, we strictly follow the MVP Passive View paradigm. Figure 4.3 shows the three main layers of the ASDM in the context of an entire SDS and of the ATRACO IE used as external knowledge source.

From the IA's perspective, as described in Sect. 2.2, the ASDM acts as a mediator between the AS and the user. Hence, the ASDM accesses the sphere via the IA. Since UPnP (ISO 2008) is one of the most common and widespread ATRACO-supported protocols, we have agreed on an UPnP-based interface between the IA and the ASDM. This interface can also be used by other external components or other types of IEs. Its definition is presented in Table 4.2 as part of Sect. 4.2.1. Before explaining the technical details required to understand the functioning of the OwlSpeak ASDM, we introduce an ATRACO specific technical concept that is used to emit information gathered by the ASDM. UPnP provides the feature of so-called "state-variables". Other components (in our case the IA) can subscribe to these variables and get notified in case of a change of state.

In ATRACO we have defined a "lastUserWill" variable that provides the latest sematic value the user triggered by uttering a specific command. Figure 4.4 shows how a user command is understood by the system and how the IA receives the

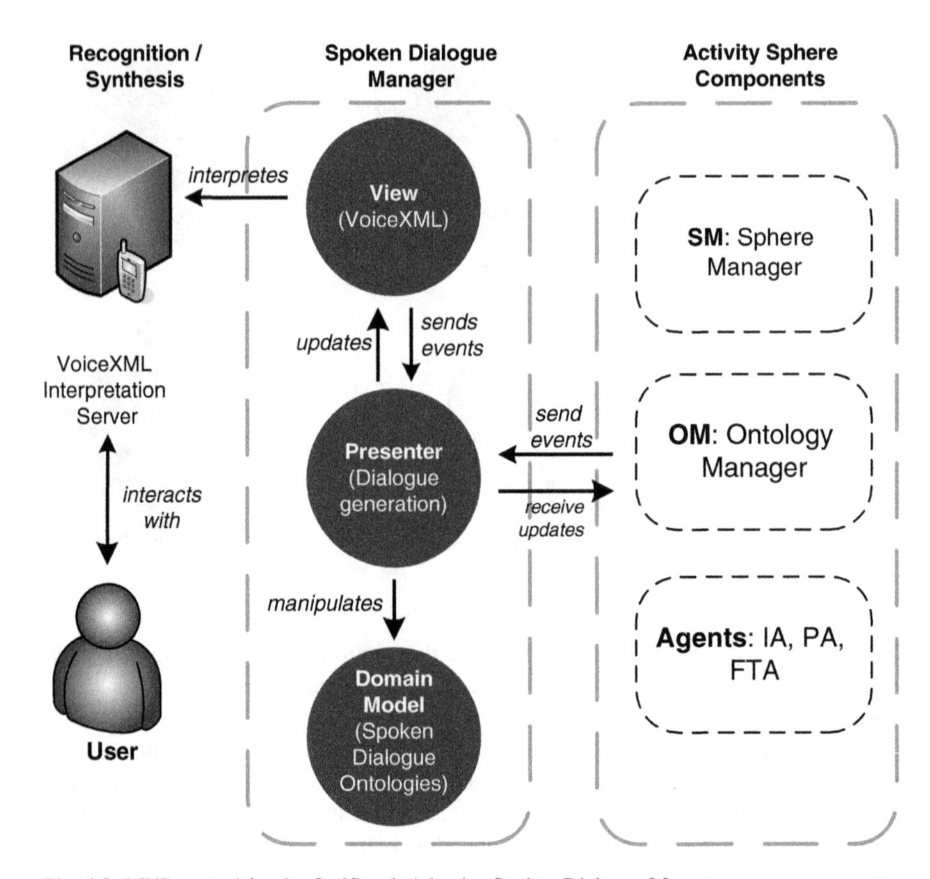

Fig. 4.3 MVP as used for the OwlSpeak Adaptive Spoken Dialogue Manager

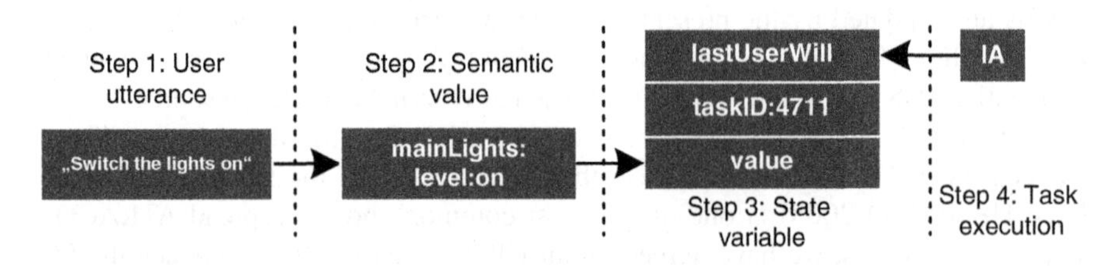

Fig. 4.4 From the user utterance to the task execution by the IA

specific information in order to redirect the command to the sphere manager that finally carries out the corresponding task. The first two steps are realised by the ASDM. Its main task is to understand the utterance and to select a semantic meaning that in turn can be understood by the ATRACO system, i.e. by the specific component that should show a reaction to the command. Step 3 is provided by the UPnP-based interface of the ASDM: the state variable provides the new semantic value for the light level ("level"), which is "on".

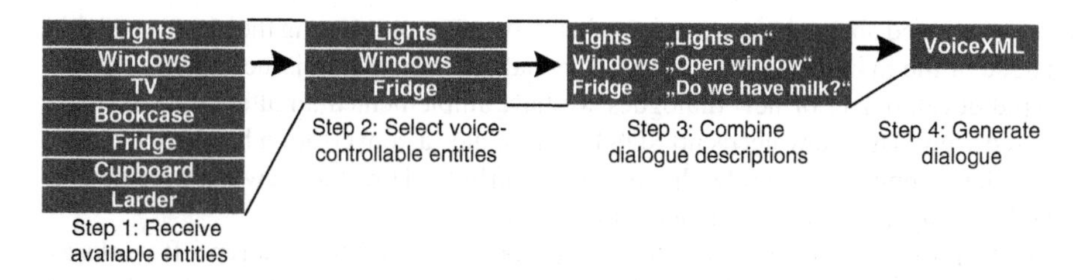

Fig. 4.5 The selection of dialogue domains by the IA and the dialogue generation by the ASDM

This value is related to the task ID of the respective task ("mainLights" in Fig. 4.4). The binding of task IDs and spoken commands is done dynamically during the initialisation of an ATRACO system and updated during the emergence of the AS. In Step 4 the IA, which is registered on the "lastUserWill" variable, is notified that a change of state happened and receives the new value of the light level for the specific task. Thus, the agent is able to redirect the user request to the sphere manager that takes care of all task execution activities within the sphere.

Figure 4.5 shows the opposite direction of the process described above: in Step 1 the IA receives a list of all entities that are capable of user interaction within the AS. Since the information related to these entities is accessible within the sphere ontology, the IA is able to decide which modality is beneficial for which entity. In Fig. 4.5 the IA selects the lights, the windows, and the fridge to be accessible via voice (Step 2). During Step 3 the ASDM is fed with the information about the selected entities and combines them into a unified dialogue description, established by a set of Spoken Dialogue Ontologies (SDOs—cf. Sect. 4.3). Finally, in Step 4 the ASDM generates a compilable and interpretable dialogue using, for example, VoiceXML. This dialogue is valid until an SDOs is modified—or until a communication with the system starts.

In combination with an IE, the ASDM forms a flexible framework that can easily be extended and adapted to different contexts and purposes by adding new or altered SDOs. These can be dynamically applied to already existing models. In OwlSpeak we use a varying number of dialogue representations as domain models. Each representation provides knowledge about both dialogue flow and state of a specific spoken conversation. Depending on contextual information, various sets of spoken dialogues can be activated or deactivated. Furthermore, it is possible to add new representations of dialogues during runtime and therefore extend the knowledgebase, i.e. the model. To implement the model, we use OWL ontologies as described in Sect. 4.3. In Sect. 4.3.3 we provide a detailed look on the capabilities of this structure and the way we use it as a meta-description for spoken dialogues.

In general, the view can be implemented using different methodologies and techniques. The main matter is that this layer must provide output that can be understood by external entities and, of course, must accept and understand input, which may be passed to the model via the presenter layer. We use VoiceXML dialogue snippets for the OwlSpeak prototype. These are generated on the fly depending on the state of the dialogue representations that are currently activated.

As mentioned above there are many other ways of implementing the view—and this is one of the main benefits of the MVP paradigm. For testing issues and to support rapid development of new dialogues we have implemented an alternative HTML-based view. This view needs no SDS but simulates a conversation between the user and the computer by simply clicking on hyperlinks. How the view is implemented and generated is explained in detail in Sect. 4.4.

The presenter layer provides the actual intelligence of the prototype. It works as a connector and translator between model and view. Implemented as a Java Servlet it generates the view by reading the model and extending its knowledge in a cyclic manner. Since the presenter plays a distinguished role for the prototype, we provide detailed information of this component in the following section.

4.2 The Presenter

We designed a layered architecture for the dialogue logic, i.e. the presenter. Three layers provide different functionalities and have different objectives so as to maximise the modularity of the prototype. This approach facilitates the reuse of the actual logic even if the view-generating layer, the Java Servlet itself, is replaced by other technologies. Such technologies may realise a direct connection to a speech development kit, e.g. to the Microsoft Speech API or to the Nuance SDK.

Furthermore, the layered approach is beneficial since the system may serve as a framework for developing new OwlSpeak-based applications apart from the management of spoken dialogues. As shown in Fig. 4.6 the three layers that compose the presenter are:

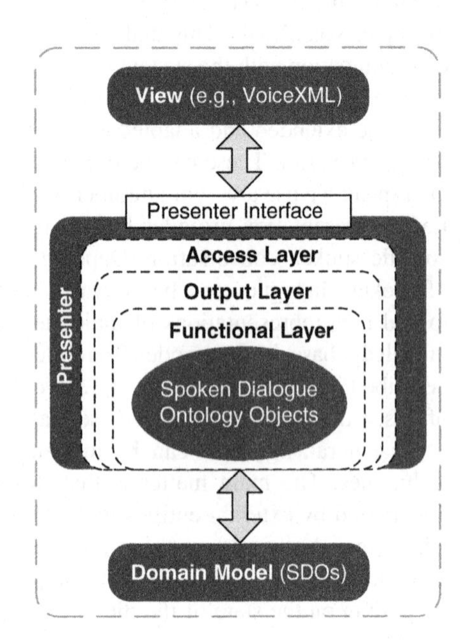

Fig. 4.6 The three layers the presenter consists of

1. The topmost level of the OwlSpeak presenter is the *access* layer. It initialises the system and constructs all objects used by the lower layers. Furthermore, this layer provides the primary connector to the view and therefore to the SDS itself. Thus, it affects the format of the view, which facilitates the process of designing alternative views that may be used by a variety of external SDSs. The parameters that are needed to correctly classify the information the user provided are managed here and (partly) passed to the lower layers as well.
2. Within the *output* layer, the structure of the generated input and output is determined. Two stacks are being prepared: one for the input the system accepts (i.e. the grammars) and one for the output the system may utter (i.e. the utterances). Therefore, the output layer utilises the functional layer to retrieve the knowledge of the model. Afterwards it passes the parsed and restructured information to the access layer.
3. The lowest prototype level is the *functional* layer. It provides all fundamental functions of the system that, for example, decide what must be uttered and what must be understood. All updates and modifications of the dialogue representation (i.e. the model) are also handled by this layer.

As defined by the MVP pattern, the presenter receives input that is provided by the view. It also sends updates to the view in order to keep the spoken interface up-to-date. The received information is passed to the output layer, which updates the input and the output stacks of the functional layer. These will then be used to generate an updated view and the process begins anew by passing newly gathered information to the access layer. The so-called SDO Objects represent and provide access to the knowledge encoded within the domain model. They compose the core of the functional layer. We utilise the OWL API to establish the programming interface to the SDOs (Horridge et al. 2007; Bechhofer et al. 2003). In the following we take a closer look at the presenter interface and explain how it can be accessed from the view and from the IE.

4.2.1 Presenter Interface

Two parts of the interface allow data exchange with the presenter: one interacts with the view and the other interacts with the environment (e.g. with the ATRACO system). The former part of the interface provides the main dialogue-enabling functionalities. Since the view consists of a VoiceXML document (see Sect. 4.4), HTTP requests are the most elegant way to send and receive data. Thus, we have implemented the presenter as a Java Servlet (cf. Sect. 4.2). This also facilitates the portability to different platforms and the accessibility within a networked context. Table 4.1 shows the three main commands and their corresponding parameters that allow the system to continually keep the actual view up-to-date and to process all user inputs by refreshing the underlying domain model, i.e. the respective SDO(s).

Table 4.1 The commands of the direct interface between view and presenter

Command	Parameters	Description
request	user (optional); event (optional); nomatchRec (optional)	This command is automatically executed after a specified phase of inactivity (i.e. if neither the user nor the system utters any word that can be aligned with the dialogue model.) It requests a new VoiceXML dialogue snipped from the presenter. Depending on the actual state of the SDO(s), it encodes new dialogue turns, that may be influenced by the state of the environment. In most cases the "request" command is parameterless. However, in order to enable the ASDM to handle multiple users, the name of the user may be submitted (e.g. user = "suki"). For the enhanced recognition capabilities of OwlSpeak two other parameters are necessary: "event" and "nomatchRec". The former parameter is used to transmit the type of event that occurred during dialogue [e.g. nomatch or noinput as described in Oshry et al. (2007)]. The latter parameter is used to transmit the user's utterance to the ASDM in case of a non-understanding that is indicated by the nomatch event. The utterance may be encoded as a WAV file. See Sect. 4.2.3.4 for further details
work	agenda; move; user (optional); speak (optional)	This command is triggered if the user utters a word or sentence that is matched by the currently activated grammar. If this happens, the presenter must update the underlying SDO(s). Afterwards it automatically executes a "request" in order to generate a new view that can be interpreted by the speech server. The mandatory parameters for this command are "agenda" and "move". The first parameter is required by the presenter logic for the location of the actual dialogue turn that has been carried out during a conversation. For single commands such as "lights on" the agenda is not necessary (it is set to a default value). For technical reasons, the value of the second parameter "move" consists of two parts: the unique name of the move (i.e. the atomic dialogue step to be carried out) and the name of the SDO it belongs to. The two important concepts of agendas and moves are presented in Sect. 4.3
error	–	If the "request" or the "work" command fails and the dialogue cannot be continued as intended, an error message of specified granularity can be passed to the view. This message may be used to inform the user or to log the unwanted situation for debugging. This behaviour can be inhibited in case errors may obfuscate novice users

Since VoiceXML does not allow Ajax-like methodologies (Garrett 2005), the view is required to regularly trigger the presenter to check for any modifications of the knowledgebase. In practice, we use a time slot of 3–7 s of inactivity between two checks. If a correct recognition or a non-understanding occurs before the 7 s elapse, the presenter is automatically accessed. In case the user utterance matched the grammar, the "work" command is executed. Thus, the newly gathered information is passed to the presenter, which adds the semantic meaning to the respective SDO. Depending on the updated knowledgebase a new dialogue snipped is generated to provide a refreshed view. The dialogue-related commands require specific parameters. Table 4.1 lists these parameters.

In case of a "request" command, the presenter logic may cope with different types of events. These events are of interest for enhancing the recognition and the understanding capabilities of the ASDM. Here, the last user utterance (encoded as a WAV file) also is a valid parameter. The utterance is submitted in case the view is realised as a VoiceXML dialogue. The "work" command encodes two parameters: the current dialogue step (i.e. move) and the current dialogue turn (i.e. agenda). This information is crucial: the presenter is not able to correctly update the knowledgebase, i.e. the SDO(s), if this information is missing. Once the presenter received the data, it updates the respective SDO(s) and calls the "request" command in order to generate an updated view. We describe the realisation of the interplay between the agendas and the moves in Sect. 4.3.

The latter part of the interface allocates a set of commands that may be used to control the ASDM itself, i.e. to activate and to deactivate SDOs or to manipulate their content. These control commands can be accessed by specified spoken commands, by HTTP requests, and via UPNP (ISO 2008). If spoken commands should be used to control OwlSpeak, both the meaning and the wording of those commands must be defined. A dedicated SDO, the system ontology, can be utilised for this purpose. The functionality of controlling the dialogue itself via spoken input relates to Behavioural Adaptation as defined in Chap. 3 and listed in Sect. 3.6. Within the framework of ATRACO, the ASDM is accessed via UPNP. Table 4.2 shows the complete set of control commands that supports UPNP-based and spoken dialogue-based control. The latter is possible if a grammar for each of the commands is defined. We have divided these commands into three categories: SDO-related, settings-related, and variable-related.

The main purpose of the SDO-related commands is to manage the set of SDOs that are used to generate the spoken dialogue. The ontologies can be reset, i.e. all a posteriori dialogue information is deleted. Thus, a specific SDO—or the entire set of SDOs—provide the dialogue models as initially defined. A single or a specific (sub-)set of ontologies may be activated or deactivated—this halts or resumes the dialogue. If an external entity influences the SDOs during run-time, all SDO(s) may be reloaded in order to generate an updated dialogue. This allows to bilaterally control the dialogue: either from the part of the IE or from the part of the user. If all ongoing spoken dialogues have to be stopped for any urgent reason, all SDO(s) can be deactivated in order to set the ASDM into a "doing nothing" mode. The settings-related commands may be used, for example, to trigger the ASDM to reload

Table 4.2 The commands of the interface between IE and/or the user and the presenter

Command	Parameters	Description
reset	onto	This command may be used to reset a specific SDO. Thus all information that has been gathered by using this SDO will be deleted. In practice this means that the dialogue will be set to the initial state. The command therefore also corresponds to restarting a conversation
resetAll	–	If this command is called, all available SDOs—active or inactive—will be reset. The ASDM would then re-initialised
disableSDO	onto	In order to enable the IE to deactivate a specific dialogue this command might be used. The dialogue will keep its current state but the presenter will not read or update any information encoded within a deactivated SDO
enableSDO	onto	In order to enable the IE to activate a specific dialogue this command may be used. As soon as the view is updated, the dialogue encoded within an activated SDO will be, optionally depending on its preconditions, included within the generation of the actual view
getActiveOntos	–	This command provides a list of all active SDOs that are currently used to generate the view
loadSDOs	–	In case a third-party component, i.e. not the presenter itself, modifies one of the SDOs this function must be called in order to trigger the presenter to re-consult the knowledgebase
stopAllDialogue	–	If, for example, due to privacy issues, all dialogues have to be stopped, this command deactivates all SDOs at once
reloadSettings	–	In case the configuration of the ASDM encoded within an XML file has been changed while the system is running this command has to be executed. If, for example, the language or the prosody of the SDS has to be changed, the system configuration has to be reloaded
setAsrMode	asrMode	This method sets the language the ASR has use to English, German, or any other language the SDS provides
setTtsMode	ttsMode	This method sets the TTS language to male or female, English or German, or any other language the SDS provides
getSleep	–	This command provides true if the system is asleep or false if it is currently active
setSleep	sleepState	The ASDM might be paused by using this command. All SDOs and settings will be kept but the system does neither understand anything nor provides any spoken output
getVariable	variable	If a component of the surrounding IE wants to get the current status of a variable encoded within a specific SDO this method may be utilised
setVariable	variable	If a component of the surrounding IE wants to set the content of a variable encoded within a specific SDO to a new value this method may be utilised

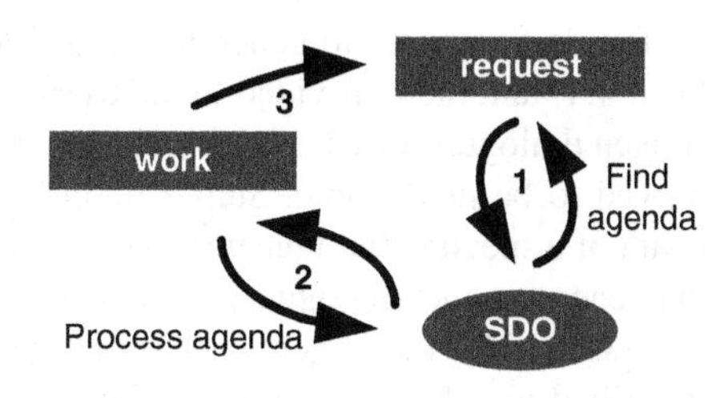

Fig. 4.7 The request-work cycle and its relation to the SDO(s)

its configuration. In case an external entity changes any configurational data, it is required to call this command. Furthermore, the ASDM is able to change the language during runtime.

Therefore, the ASR and the TTS have to be set to a specific language that must be provided by the SDS (the current prototype allows English and German). If the spoken interaction has to be halted for any reason, the ASDM may be set asleep. Within this state all dialogues are still active. However, the system does neither understand anything nor utters a single word. A third class of commands is directly related to the content of the SDOs: variables, which are specific fields encoded as part of the SDOs, can be modified during runtime. Variables are useful since they allow, for example, to modify a specific system utterance during an ongoing dialogue. They can also be used to define conditional dialogue flows that are triggered from outside, i.e. by the environment. The concept of variables realises the event-based information exchange as depicted in Fig. 4.4 (see Sect. 4.3). In the following section we discuss the basic functionality of the OwlSpeak ASDM that renders the user interface.

4.2.2 Basic Functionality of the Presenter

The basic functionality of the OwlSpeak ASDM is to generate spoken dialogues consisting of dialogue turns or conversational acts that should subsequently be carried out. Examples of such dialogues are booking tasks or, as described in the application scenario, the exchanges initialised by the system (see Table 3.5). Utilising an SDO to define such information retrieval dialogues is similar to utilising other descriptive approaches such as VoiceXML. This specific type of dialogue is system-directed. Hence, the system queries the user and/or is able to understand specific questions uttered by the user (e.g. "what can I do"). Figure 4.7 illustrates how the two basic methods "request" and "work" are geared with the knowledgebase, the SDO(s). Three steps are depicted: Step 1 is carried out if the ASDM checks whether the knowledgebase has been modified, i.e. if any of the dialogue domains experienced a modification. For this purpose, the most urgent agenda (in other words, the most urgent dialogue turn) and all control commands that are active are detected and combined to a new dialogue description. Thus, the dialogue logic's objective is to *identify the correct agenda*.

The data required to generate a dialogue description are passed to the view (see Sect. 4.4) and then provided to the user. Step 1 is carried out as long as there is no spoken dialogue activity. During the test runs "listening phases" from 3 to 7 s have proved to be practicable. Step 2 describes the update of the knowledgebase as a result of a specific dialogue turn. This "work" method adds the semantic knowledge that encodes the dialogue progress to the SDO(s). Thus, a specific *agenda is processed.* The fundamental algorithms used to provide these functionalities are described as UML Activity diagrams in Appendix D (Figs. D.1 and D.2). After the knowledgebase has been updated, the "work" method automatically calls the "request" function (Step 3). At this point, the cycle starts again by carrying out Step 1. The rigid distinction between logic and knowledgebase, i.e. SDOs, enables multitasking and dialogue focus switching. Referring to Sect. 3.6.1 and to the application scenario we notice that the proposed basic functionality is crucial to (partly) realise Behavioural Adaptation, Dialogue Strategy Adaptation, Device Adaptation, Event Adaptation, and Task Adaptation. In the following we present technical solutions required to complete the functionality of the OwlSpeak ASDM that goes beyond pure descriptive approaches to spoken dialogue management.

4.2.3 Enhanced Spoken Dialogue Management for Intelligent Environments

Aside from the basic functionality required to provide a dialogue flow covering multiple tasks it is necessary to introduce several concepts that complete the adaptive aspects of the proposed prototype. In Sect. 4.2.2 we describe how dialogues that are executed serially can be realised. Such dialogues may also be halted while other dialogues are activated, which is crucial to allow multiple dialogue topics. However, this basic approach is not fully appropriate to cover all IE requirements: the dialogues that run in parallel may have to be executed concurrently or cause conflicts in case the dialogue domains overlap. As part of the presented work, several methodologies have been implemented to allow parallel execution of dialogue topics. These are described in the following.

4.2.3.1 Concurrent Dialogues

As conceptually described in Sect. 3.6.1 the ASDM is able to activate several dialogues in parallel in order to provide interfaces to different devices and services populating an IE such as the ATRACO system. Subsequent to each "request-work" cycle, the ASDM updates the set of available dialogue descriptions (i.e. SDOs). During a second step all valid agendas are extracted in order to establish a list of possible utterances and grammars. Since it is not practical to utter more than one utterance at a time, the ASDM selects an utterance that is related to the agenda with the highest priority. If this utterance belongs to an exchange, the corresponding

Table 4.3 A dialogue snipped that might occur during an interactive situation

Speaker	Utterance
Suki	Hello Julia!
Julia	Hi Suki!
	Do you want to start preparing the dinner now?
Suki	Switch the lights on!
Julia	Do you want to start preparing the dinner now?
Suki	Yes!

grammars are added to the list of grammars. If the utterance belongs to a system turn, the ASDM immediately begins to generate the dialogue (in VoiceXML) that must be interpreted by the SDS. While utterances can only be serially expressed, grammars for different topics can be activated in parallel. This is actually one of the most important adaptation-enabling features of the ASDM. The dialogue logic automatically adds grammars that the system always must understand to the dialogue generation process. These grammars are tagged as so-called "respawns" (see Fig. 4.14). This allows for dialogue situations as presented in Table 4.3.

The exchange that the ASDM must carry out asks for information that is related to the dinner preparation task while the user answers with a command related to lights control. Such behaviour is realised by the "respawn" concept. The grammar for the controlling important devices and services is always added to the generated exchanges and user turns. Thus, more than one device or service may be controlled by the user in parallel. Furthermore, each of the concurrent dialogues may be deactivated during runtime. Accordingly, in case the device or service is not available anymore, the dialogue may easily be removed. This concept was extensively evaluated as part of the social ATRACO evaluation (van Helvert et al. 2009) with the dialogue domains described in Appendix A. If the dialogue domains (partly or completely) overlap, it is not possible to handle them as concurrent dialogues. In the following section we illustrate a solution of the issue of dialogue conflicts and ambiguities in ASDM.

4.2.3.2 Conflicting Dialogues

Since we allow the provision of more than one dialogue domain in parallel, we also allow the provision of grammars that may partly or even fully overlap. For example, if two lights (a table lamp and a ceiling light) are part of an IE (e.g. of an ATRACO Activity Sphere), the ASDM generates a dialogue that provides similar grammars (e.g. "lights on" or "lights off") combined with different semantics (e.g. "table_lamp_on" and "ceiling_light_on"). Figure 4.8 shows three overlapping domains: the table lamp, the ceiling lights, and a service that provides control of all available lights. Several commands are also domain specific. These allow the user to directly take control of a device.

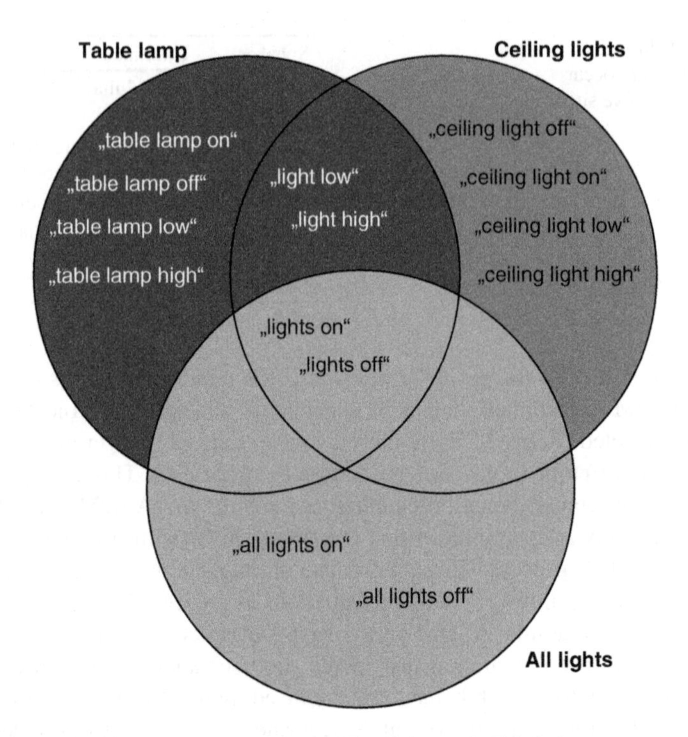

Fig. 4.8 Overlapping domains (a table lamp, ceiling lights, and all lights)

Table 4.4 A dialogue snipped showing the conflict resolution capabilities of the ASDM

Speaker	Utterance
Suki	Lights on!
Julia	Which one? *Table lamp*, *ceiling lights* or *all lights*?
Suki	Ceiling lights.

The grammars "lights on" and "lights off" are not unique: all dialogue domains provide this command. This means that the ASDM cannot automatically emit an event to the respective device in case the user utters "lights on". Such conflicts are detected during runtime and are solved by initiating a spoken dialogue for conflict resolution. For this purpose, the system generates an exchange to ask the user to which domain he refers to. In order to generate such conflict resolution dialogues on the fly, we introduced the concept of *DomainNames*. Each dialogue domain provides a unique name that has to be human-understandable. This name is used by the ASDM to automatically generate an exchange in order to solve the conflict. The system uses the DomainName to compose both the question and the grammar, i.e. the possible user answer. Table 4.4 shows a dialogue snipped describing a conflicting situation.

As depicted in Fig. 4.8, such a situation may occur if the domains are combined. The name of the domains (in italics) may be utilised by the user to answer. A main benefit of this approach is that the conflict resolution is realised at the lowest level of the dialogue: at the grammar level. Only if the user utters a command that matches one of the conflicting grammars indicated in Fig. 4.8, the conflict resolution dialogue will be initiated. Furthermore, the ASDM is able to detect all domains that are involved in the conflict. In case the user utters “lights on” in our example all three domains will be taken into account by the system. Analogously, in case of uttering “light low” only the table lamp and the ceiling lights are included as possible options that may be controlled since the service for controlling all lights does not provide a grammar for dimming. A drawback of this approach is that unique and human-understandable names for each of the domains are a prerequisite. It may be possible to ask the user for a proper name if a new device is introduced in the IE to compensate for this disadvantage.

A further issue that relates to SDSs in general is the scalability of the conflict resolution dialogues. In case the environment provides hundreds of lights that can be controlled separately, a “lights on” command would force the SDS to list all friendly names of all the devices. The user would not be able to perceive them all. However, during the evaluation of the ATRACO system we have noticed that usually there are not more than seven devices per room that may provide similar grammars and therefore lead to conflicts. According to Miller (1956) and to our own experience gained during the evaluation, we assume that seven possible alternatives are the maximum number of device names the ASDM may provide when generating a conflict resolution dialogue. In the following we take a closer look at how the ASDM is able to judge which agenda, i.e. which conversational act must be carried out next—this decision is a further important multitasking enabling aspect to be taken into account.

4.2.3.3 Prioritisation of Dialogues

For the provision of meaningful multitasking, methods are required for prioritising the different dialogue turns. Without such functionality, the ASDM would not be able to decide which turn to be carried out at one particular time. A straightforward way to select a turn or task is to allocate numeral priorities. As shown in Fig. 4.14 each agenda provides a data field that stores a value between 1 and 100 in order to indicate its urgency. This value can be defined during design time or modified by an external entity during run-time. In the following we call this feature *static* prioritisation. When a new turn is added to an SDO, the system automatically allocates an initial value if the dialogue designer has not selected a specific one.

However, more sophisticated approaches are conceivable to enable efficient and comfortable multitasking over different domains. As described in Sect. 3.3 there are many options how an ASDM could decide which task, i.e. which conversational act or which dialogue to be selected. To name but a few: static prioritisations, timer-based system, semantic approaches, and logical decision rules may be utilised. We have decided to implement an extended version of a timer-based system that in the following is referred to as *dynamic* prioritisation. This kind of prioritisation takes the static priority of an agenda (a dialogue turn) into account and increases it turn-wise. To this end we take the current priority of a turn and multiply this value by a fractional part of the initial (static) priority. The dynamic prioritisation is therefore described as:

$$P_{new} = P_{init} + (t_{now} - t_{init}) * (\frac{P_{init}}{10}) \quad (4.1)$$

Where P_{new} is the newly computed priority of a turn, P_{init} is the priority the turn had when the dialogue was initialised, t_{init} is the time (measured in dialogue turns) the dialogue was started, and t_{now} is the actual time (in dialogue turns). The proposed formula enables the ASDM to allocate more time to a probably ongoing dialogue. Therefore it does not have to interrupt a conversation in order to utter a warning or notification. By default, all turns are set to an initial, predefined priority between 1 and 100. This enables the ASDM to sort the different dialogue turns that are encoded in the SDOs. However, if the dialogue designer allocates a very low initial priority to a specific warning, the ASDM is able to dynamically raise it until the warning's priority, i.e. its urgency is high enough to interrupt an ongoing dialogue. In case the dialogue has already been terminated there is no need to suppress it any more. A benefit of the proposed dynamic prioritisation is that the user gains a trade-off between the growing urgency of a dialogue during a specified period of time and the change that a topic switching could be avoided.

The purpose of our approach is to avoid dialogue interruptions as best as possible. Under specific conditions, however, such interruptions may also be beneficial: for two semantically related dialogues, the user may find it useful if a sub-dialogue suppresses the ongoing dialogue. Within the context of our scenario, an example is a sub-dialogue about fish preparation helping the user in case he decides to have fish for dinner (cf. Sect. 3.5). To realise such a mechanism, the ASDM must be able to detect semantic relations between a user utterance and a dialogue: such issues can partly be solved by utilising semantic knowledgebases as we do for the adaptive recognition approach that is presented in the following.

4.2.3.4 Adaptive Understanding

In the previous paragraphs we have presented methods required to realise and to improve multitasking over different dialogue domains. These methods have been implemented as part of our prototype. In this paragraph we focus on techniques for adaptive understanding. We present the technical implementation of three approaches based on the ideas introduced in Sect. 3.6.3: *keyword-based*, *blurred-keyword*, and *semantic-keyword*. To this end we have modified the generation of the view (i.e. of the VoiceXML documents) to allow the transfer of the recorded user utterances (after a non-understanding occurred) to an external recogniser. For this purpose we use the Microsoft Speech API (MS SAPI). This API is able to utilise a grammar (i.e. a list of keywords) or to perform OOV recognition based on the MS SAPI language model. The transfer of the user input to the recogniser is performed by the built-in VoiceXML variable "application.lastresult$.recording",

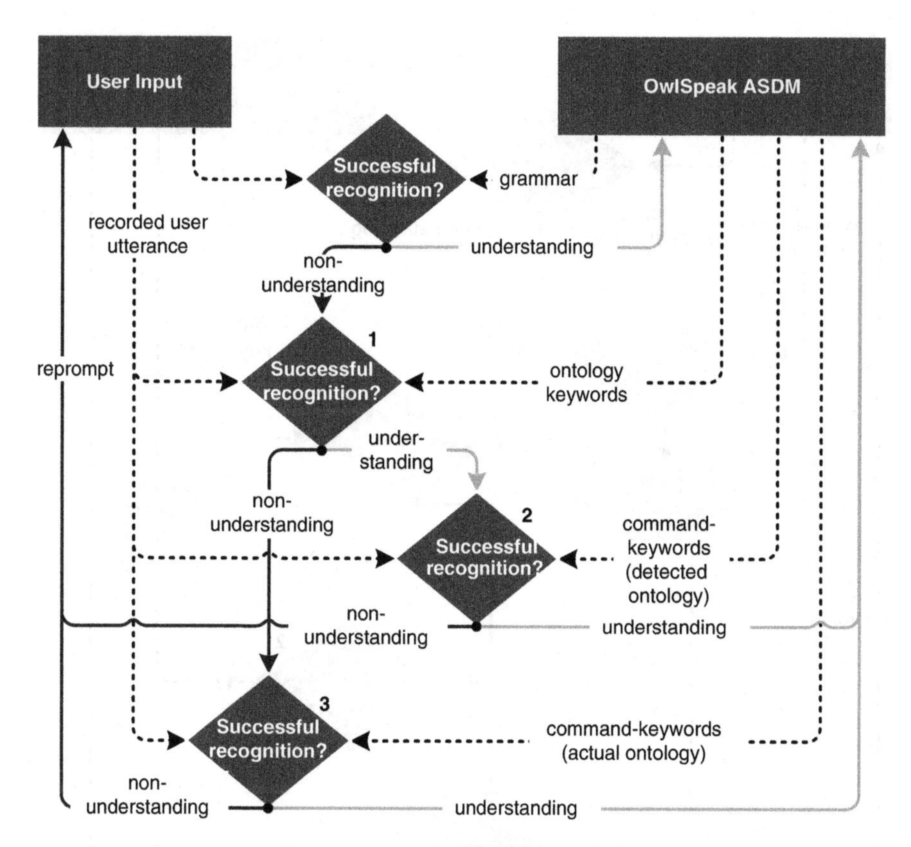

Fig. 4.9 Flow-diagram illustrating the keyword-based approach. The *dark arrows* refer to non-understandings leading to the default system behaviour. The *light arrows* symbolise an understanding that finally causes a confirmation question. The *dotted lines* show the data flow

which is passed as a wav-file to the external recogniser. Upon successful analysis of the utterance the result is passed back to the OwlSpeak ASDM, which reacts appropriately by modifying the dialogue model and generating a new view. Figure 4.9 presents a flow-diagram of the *keyword-based* approach.

It detects key utterances that are related to a specific dialogue model. In case the system fails in recognising an input such as "lights *background noise* on" in the first attempt, this approach allows to detect the command "lights on" correctly. Therefore, the approach detects "light", which is provided as an SDO-related keyword by the domain-specific dialogue model (Fig. 4.9(1)). Each SDO provides a list of keywords that significantly relates to the domain the spoken dialogue refers to. In this example useful keywords are light, lights, and illumination. Afterwards, a list of significant utterances, i.e. command-keywords for the specific domain are used to detect the word "on" (Fig. 4.9(2)). In case of a successful detection, a confirmation dialogue is automatically generated. If the system does not detect a matching input

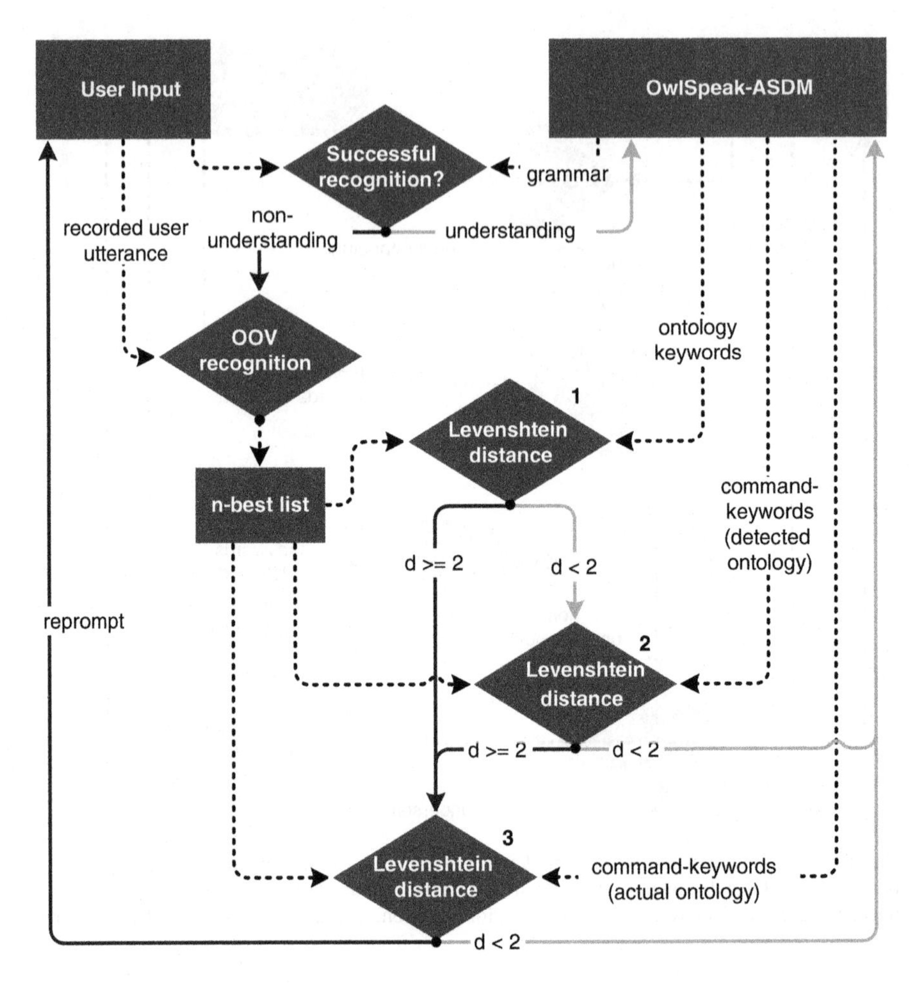

Fig. 4.10 Flow-diagram illustrating the Levenshtein-based blurred-keyword approach. The *dark arrows* refer to non-understandings leading to the default system behaviour. The *light arrows* symbolise an understanding that finally causes a confirmation question. The *dotted lines* show the data flow

a third analysis is performed using the command-keywords of the last ontology (the last dialogue) that was actually involved in the conversation (Fig. 4.9(3)). Especially for a more complex dialogue that, for example, performs the recipe selection as part of the application scenario (Sect. 3.5) this analysis is useful: the user does usually not indicate to which dialogue domain his utterances belong to. He would not ask "*Recipe selection*, do we have enough butter?" However, he would (correctly) assume that the system is aware of the actual dialogue topic, i.e. the actual dialogue domain.

Figure 4.10 describes the *blurred-keyword* approach utilising the Levenshtein-Distance. In case of a non-understanding detected by the regular recogniser, an OOV recognition is performed. The result, a n-best list, is used to pair-wise calculate the Levenshtein distance for the same keywords we have already introduced as part of the keyword-based approach. Again, during a first step (Fig. 4.10(1)) the domain-related keywords are utilised. Usually German homophones have a maximum Levenshtein distance of one. Hence, we accept a distance that is lower than two to indicate an understanding. During a second step (Fig. 4.10(2)) this calculation is repeated utilising the command-specific keywords. In case of a further understanding we continue with a confirmation dialogue similar to the keyword-based approach.

Analogously, we perform a third analysis of the user's input. Here, we calculate the distance of the command-related keywords of the last ontology that has been involved in dialogue generation and the n-best list (Fig. 4.10(3)). The default behaviour of the system is invoked in case the distance of this last analysis is greater or equals two: the system repeats the last prompt (if a question must be answered) or it behaves passively and waits for a new user input. The main difference to the *keyword-based* approach is that the ASDM covers more utterances with the same keywords. Even misunderstood input may be detected since the *blurred* keywords may also correlate with noisy input: the Levenshtein-Distance is utilised to perform error-correction on the dialogue management level. However, since a *blurred* detection may also be more error-prone than a grammar-based recognition all three strategies apply a confirmation turn in case of a supposed understanding as depicted in Fig. 4.10 (cf. Sect. 3.5.2).

A completely different problem is a user input consisting of words that are not covered by the grammar and therefore cannot be detected by the *blurred-keyword* or the *keyword-based* approaches. Figure 4.11 depicts the *semantic-keyword* mechanism incorporating a semantic knowledgebase to analyse the user's intention and to cope with this issue. If the regular recogniser detects a non-understanding, the n-best list provided by the OOV recognition is semantically analysed using the GermaNet API (Fig. 4.11(1)). The system uses the hyponyms of all SDO-related keywords and compares them pair-wise in order to find semantic similarities. As mentioned above, all SDOs provide domain-related keywords. In other words, the system is able to detect "lamp" or "torchiere" for the dialogue domain keyword "ceiling light". This broadens the domain coverage of the keywords and allows for colloquialisms. In case of a positive match, the system checks whether a command-related keyword is part of the n-best list (Fig. 4.11(2)). In case of a non-understanding, the system proceeds with the dialogue and ignores the last user input. The following comparison is processed analogously to the keyword-based approach (Fig. 4.11(3)). For the two steps in the process we disregarded from an optional comparison based on hyponyms since GermaNet does not provide verbs and only a few adverbs. However, if other knowledgebases such as WordNet are utilised, Step 2 and 3 can also be semantically processed. Again, in case of a positive match, the system asks the user to verify the utterance. If the user confirms, the ASDM carries out the command or incorporates the newly gathered information into the SDOs.

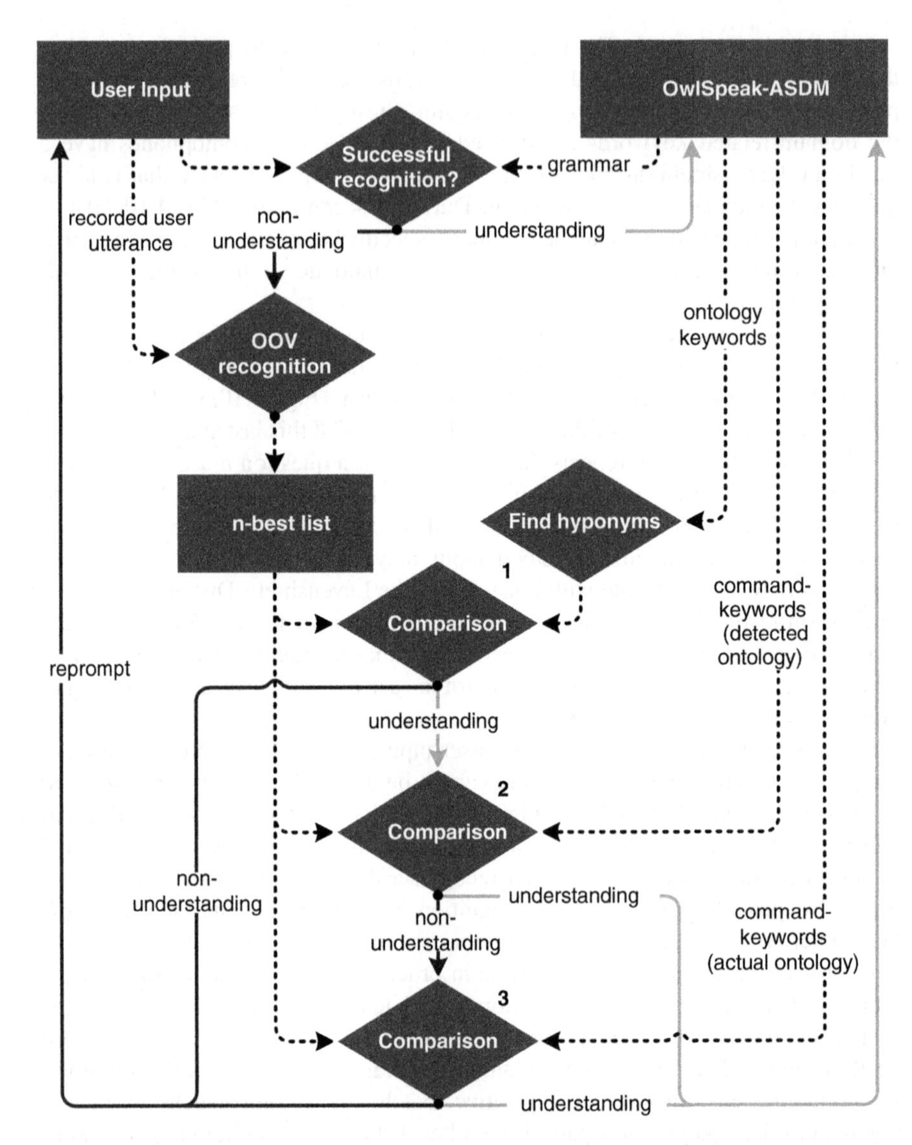

Fig. 4.11 Flow-diagram illustrating the semantic analysis using the GermaNet-based semantic-keyword approach. The *dark arrows* refer to non-understandings leading to the default system behaviour. The *light arrows* symbolise an understanding that finally causes a confirmation question. The *dotted lines* show the data flow

We argue that an OwlSpeak-based SDS benefits from the presented approaches when it comes to spontaneous and intuitive user–system communication. In particular under the open-domain conditions as described in the application scenario, spoken commands occur spontaneously even during an ongoing dialogue that may be related to a different topic. Such use cases require not only intuitive handling but also more sophisticated methods to *understand* what the user is talking about. Hence, we emphasise the importance of the capability to influence the recognition behaviour of the SDS by the ASDM. We have defined these methods as part of the SDS-centred *Speech Adaptation* (cf. Sect. 3.2). After presenting the enhanced functionalities of the presenter we investigate the knowledgebase that is used as a foundation of the dialogue logic. Thus, in the following section we present the domain model and describe how the presenter utilises these dialogue domain descriptions to generate a meaningful output that may be rendered by the view layer and interpreted by the SDS.

4.3 The Domain Model

The underlying knowledgebase of OwlSpeak is modelled using OWL ontologies, so-called SDOs. We utilised the tree-shaped structure of OWL to develop a theoretical foundation of the ASDM. A defined set of classes is used to arrange the data-bearing individuals (McGuinness and van Harmelen 2004). This enables the system to programmatically handle the dialogue models. Our proposed set of classes is depicted in Fig. 4.12. The root of the knowledgebase is *DialogueDomain*, which consists the two subclasses *Speech* and *State*. The class DialogueDomain also provides two important data fields realised as so-called OWL annotation properties: domainName and domainKeyword.

The data field domainName may be used to define a human-understandable name that may be used to generate, for example, conflict resolution dialogues. It is also possible to define alternative names in order to allow for providing a more detailed description of the dialogue, the device, or the service. The data field domainKeyword may be used to define the keywords to be utilised for advanced understanding strategies as described in Sect. 4.2.3.4. A keyword does not have to consist of a single word—it is possible to define expressions such as "wholefood cookbook". Furthermore, it is possible to define more than one keyword per DialogueDomain. It is not a trivial task to define the keywords since they may have a heavy influence on the spoken interaction.

We divide the SDO into the two main branches of Speech and State since we distinguish between knowledge that corresponds to the static dialogue structure and knowledge that dynamically corresponds to the state of the actual dialogue. Hence, all subclasses of Speech describe the basic blocks used to compose a dialogue. The most complex structures realised as part of the class Speech are *dialogue steps*. During the conversation a dialogue step is a specific input by the user or a specific output uttered by the system. A dialogue step always provides a specific meaning, i.e. a semantic value. More complex dialogue structures such a dialogue turns, consisting of various steps, are composed as part of the class State. This inherent part of the SDO describes the dynamic aspects that undergo modifications while the dialogue progresses.

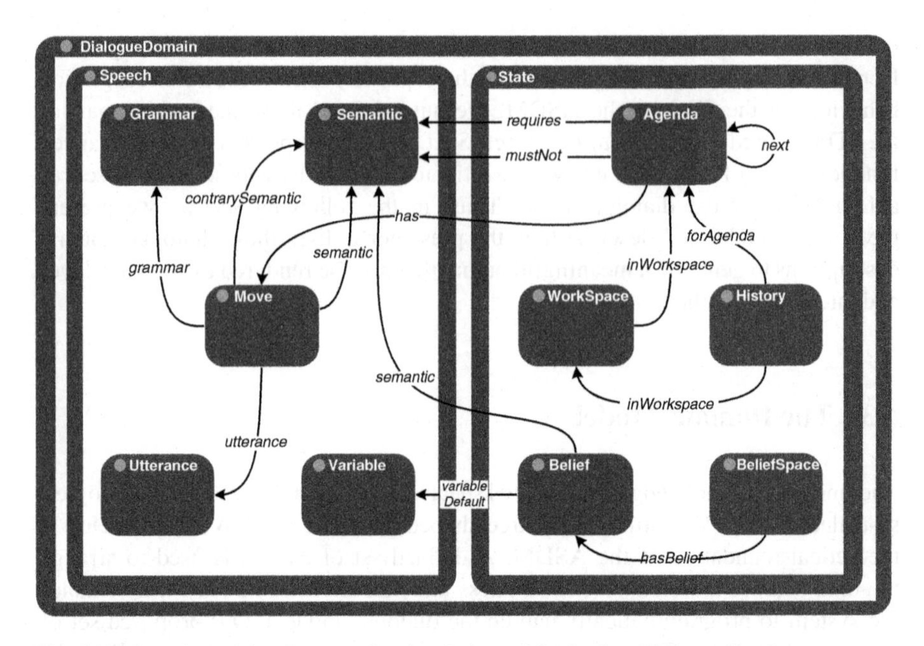

Fig. 4.12 Overview of the classes and main relations of the Spoken Dialogue Ontology

Figure 4.12 shows an overview of all classes populating the SDO and the relations interlinking them. Some naming conventions must be defined since the following sections provide an in-depth description of the OWL spoken dialogue domain model, the SDO. Classes such as Move and Grammar always start with a capital letter. Individuals, the data-bearing objects, always belong to a specific class. They are marked with the name of the respective class after a "_" and start with a lower case (e.g. question_Move). Logical relations and data fields are italicised (e.g. *requires*).

4.3.1 Static Knowledge

To develop an SDS two basic aspects have to be interrelated in a meaningful way: on one hand statements by system and on the other hand statements by the user that must be understood by the system. The relationship of the two aspects is handled similarly during a human–human dialogue. In the following we use the term "utterance" for a statement uttered by the system and the term "grammar" for a statement the system should be able to understand. Figure 4.13 shows the subclasses of Speech, which describe the static part of the SDO. As illustrated, two sets of character strings are used: one as part of the class Grammar and one as part of the class Utterance. The utterances can be passed directly to a corresponding TTS that conveys them into speech.

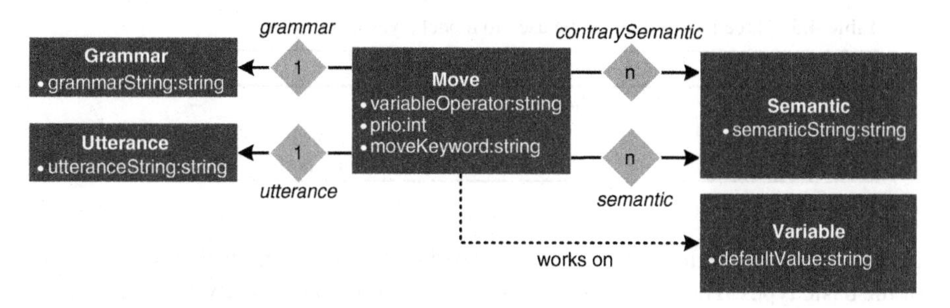

Fig. 4.13 The classes and relations of the static part of a Spoken Dialogue Ontology

The grammars define the user input that can be understood by the SDS. Therefore, a standardised format—interpretable by different SDSs—must be chosen. Since the prototype uses VoiceXML to render the view, all VoiceXML compatible formats are appropriate [e.g. SRGS (Brown et al. 2004) or JSGF (Hunt 2000)]. The implemented prototype utilises Nuance GSL (Nuance 2008) to describe the grammars. A typical grammar is "[lights light (lights on) (light on) (switch the light on)]".

Since a grammar that encodes a positive answer such as "yes" may have varying meanings depending on the actual question, it is required to combine the grammars (and the utterances) with semantic values. The SDO class Semantics aggregates these values. All semantic values are, within a specific context, meaningful to the ASDM and/or the user. Figure 4.13 shows two relations between the class Move and the class Semantic that are both used for this purpose. The class Semantic itself stores a semantic meaning or a specific value that is understandable and interpretable by the system or by an external entity, which is connected to the ASDM (e.g. "user_is(hungry)").

The three basic classes Utterance, Grammar, and Semantic establish the main static knowledgebase of OwlSpeak. To complete the concept, a fourth type was added to the ontology framework. It can be used to influence the static content of the SDO during runtime: the class Variable. This class contains a semantic value similar to the Semantic class. However, in this case the value can be influenced by a set of logical operations. To exemplify this feature, we may imagine an utterance such as "Hello x". Using the concept of variables, we can set, for example, "x = Suki" during runtime. Thus, it is not required to define a greeting for all names of all users. A further difference between semantics and variables is that variables can be influenced from outside the ASDM, e.g. from the ATRACO IE. Contrary to this, semantics can only be read from outside but not set nor modified. The ASDM rather utilises semantics to express the dialogue flow and its actual state.

Since these four types only provide character strings, further concepts are required in order to express the actual dialogue functionality. For this reason, we introduce the class Move. It acts as a container element that interconnects the four basic types Utterance, Grammar, Semantic, and Variable. A Move relates to

Table 4.5 Three moves that can be used to model a yes/no question

Move	Utterance	Grammar	Semantic
question_Move	"Are you hungry?"	–	–
answer1_Move	–	[yes (yes i am)]	answer(true)
answer2_Move	–	[no (no i am not)]	answer(false)

a dialogue step as defined above and is shaped by four relations each allocating one of the basic types. The *grammar* and the *utterance* relations directly point at exactly one grammar and one utterance individual, respectively. A move either encapsulates one grammar or one utterance using these relations.

We assume a single statement cannot be uttered by both the system and the user simultaneously. Analogously, the *semantic* relation points at a semantic individual. A move may have more than one semantic meaning—if the relation points at two semantics they are combined using a logical "and". The *contrarySemantic* relation points at all semantics that should be suppressed by a specific move. This is the inverse to the *semantic* relation: in case a specific move is processed, the semantics it refers to are not valid anymore. Therefore, they will be dropped from the memory of the ASDM—its belief state (cf. Sect. 4.3.2).

The class Move also provides an integer value that describes the priority of the move individual (*prio*). How this priority is used in practice is explained in detail in Sect. 4.2.3.3. The *varibleOperator* field may be used to store specific commands that alter a variable if the move has been processed. For this purpose, a logical parser is included as part of our prototype in order to evaluate logical expressions (see Sect. 4.3.2). Figure 4.13 depicts a dotted line between the classes Move and Variable expressing that a specific move may influence the value of a variable. Table 4.5 shows a set of three moves that may be used to model a yes/no question.

In this case, question_Move itself would not need a semantic value. The question is always combined with at least one possible user answer providing a grammar, which assigns a semantic value itself. Thus, the question_Move solely consists of the utterance "Are you hungry?". The two answer moves differ regarding their grammar and semantics. In OwlSpeak we use moves and sets of moves to describe both the conversational acts of the users and of the system (i.e. the IE).

Therefore, it is possible to utilise the *Speech* branch of the SDO to define the entire static knowledge that is needed for a specific dialogue domain. There is no conversational structure or dialogue flow defined yet. The static knowledge including all grammars, utterances, variables, and semantics may be collected within a universal SDO and inherited into other ontologies in order to be reused. In the following we present the *State* branch of the SDO. It encodes the dialogue flow, which is typically unique for a specific conversation. Our model combines the dialogue flow with the actual state of a probably ongoing dialogue. Hence, we speak of the *dynamic* part of the knowledgebase.

4.3.2 Dynamic Knowledge

In contrast to the static knowledge explained in the previous section, this part of the SDO, the *State* part, usually is unique for a specific dialogue domain. This means that our model not only defines the static resources such as the grammar to express what a conversation composes but also addresses *the flow and the state* of the dialogue. The class Semantic stores a specific semantic meaning. However, this meaning must be transformed into conversational knowledge before the ASDM may utilise it. For this purpose, the class Belief provides its own *semantic* relation that is used to link semantics that are evaluated to be valid. A semantic is treated as valid if the corresponding move has been processed, i.e. if the knowledge that represents this move is shared between user and system.

Thereto the class Move provides the relation *semantic*: if a specific move is carried out as part of a conversational act, a new belief individual is added to the SDO referring to the semantic individual representing the actual knowledge. For the moves that provide knowledge that is contrary to already shared knowledge, the contrary semantic value can also be deleted from the shared knowledge stored as a belief individual. An example describing contrary knowledge is that *a user cannot be hungry and not hungry at the same time*. If the user utters that he is *not hungry*, a semantic individual that in the past has been added to the shared knowledge, providing the information that the user is *hungry* must be deleted. Thereto the class Move provides the *contrarySematic* relation. Variables are treated analogously: if a variable is evaluated as true and/or a new value of the variable must be stored, a new belief individual is applied. This individual references the variable and associates it with a new value. The default value keeps unchanged during this process to allow the re-initialisation of the dialogue.

One of the most important concepts encoded within the State part is the Agenda class. In OwlSpeak, agendas are utilised to define the dialogue flow: they encode both pre- and postconditions and links to agendas that may subsequently be carried out. The example depicted in Table 4.5 shows three moves that in practice are pooled within a single agenda. Thus, we use agendas to define specific dialogue turns, i.e. conversational acts that consist of more than one dialogue step. The role of the class Agenda is twofold: on one hand it aggregates the static dialogue information and on the other hand it defines the flow of the dialogue: individuals of the Agenda class may provide references to other agendas using the *next* relation. These agendas may then be subsequently carried out if specific conditions, the dialogue turn may require, are met.

Figure 4.14 illustrates all relations and classes of the dynamic part. The ASDM only accesses and interprets (i.e. generates a view from) the linked agendas if both *requires* and *mustNot* relations are evaluated as true. The former relation defines that a specific semantic has already been shared with the user. Respectively, the latter relation defines that a specific semantic individual must not be referenced from a belief individual using the *semantic* relation. Since both relations may not only point at a single semantic individual but at sets of semantics these two relations

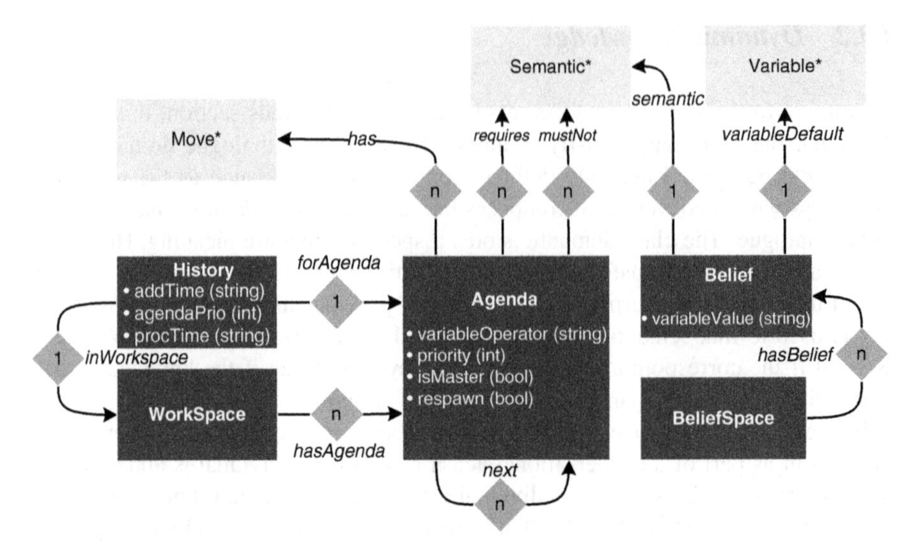

Fig. 4.14 The classes and relations of the dynamic part of a Spoken Dialogue Ontology. The related static classes are marked with a "*"

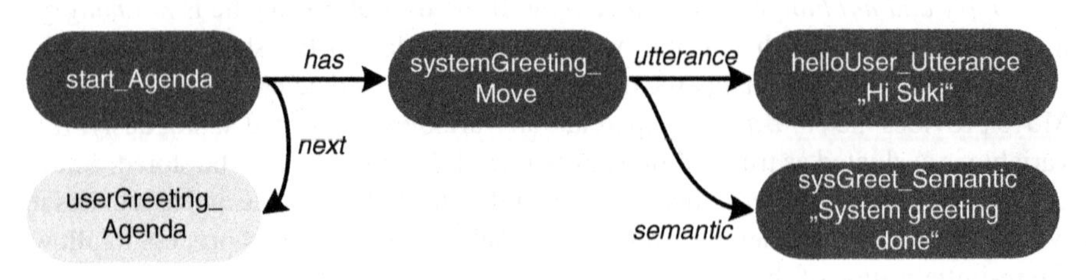

Fig. 4.15 An agenda that realises a system turn

may be used to express a huge number of logical dependencies that may occur as part of a spoken dialogue. Furthermore, an agenda individual encapsulates one or more than one move (i.e. individuals of the Move class).

We divide conversational acts into user turns, system turns, exchanges, and clarifications (that are realised as exchanges) as explained in Sect. 3.5.2. The Agenda class provides the *has* relation to reference moves. Depending on what types of moves are referenced, the system generates an appropriate conversational act. A system turn is typically represented by an agenda. It only references moves that provide an utterance, i.e. a statement that may be uttered by the system. Figure 4.15 depicts the individuals required to model such a conversational act: the start_Agenda *has* the individual systemGreeting_Move.

This move provides the two relations *utterance* and *semantic* pointing at the utterance "hi Suki" and at a corresponding semantic individual sysGreet_Semantic that may encode an additional string value. Notably, the *has* relation may be used to point at more than one move. Concerning system turns, moves may be uttered in

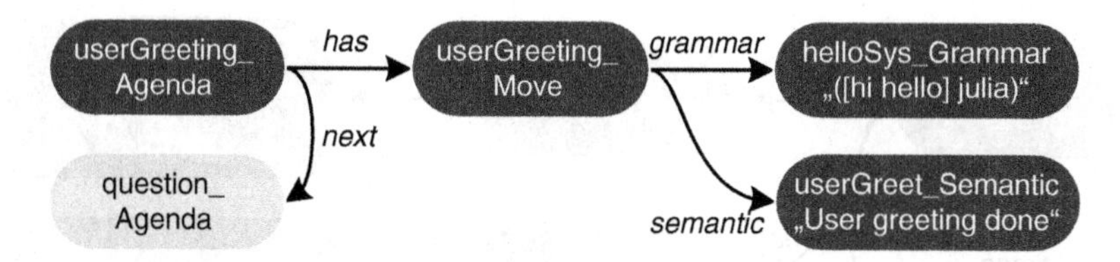

Fig. 4.16 An agenda that realises a user turn

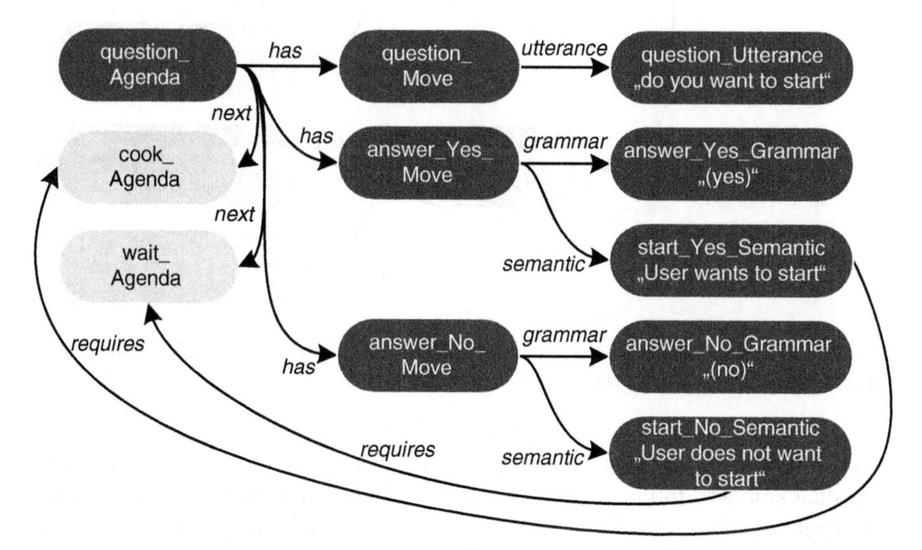

Fig. 4.17 An agenda that realises an exchange

a predefined order since they contain a *prio* value (cf. Fig. 4.13). The *next* relation links to a subsequent agenda: a user turn that is realised by the userGreeting_Agenda that is explained in the following paragraph. In contrast to system turns, user turns are represented by agendas providing moves that encode grammars, i.e. statements to be understood by the system. Figure 4.16 illustrates the individuals required to establish such a user turn. The userGreeting_Agenda references a move that consists of a grammar describing what the system is able to understand and a corresponding semantic value (userGreeting_Semantic). If no input on part of the user or no modification of an SDO occurs, the ASDM generates such a user turn in a cyclic manner, i.e. the system remains in a *listening mode*. Again the agenda provides a link to a subsequent agenda that may be processed if the semantic individual userGreeting_Semantic is referenced from a belief individual, i.e. is part of the shared knowledge.

Analogously, Fig. 4.17 depicts an exchange, which is formed by an agenda that provides both moves containing grammars and utterances. In practice, the utterances relate to questions and the grammars relate to possible answers to these questions. Depending on the user's answer one of the answer moves is processed

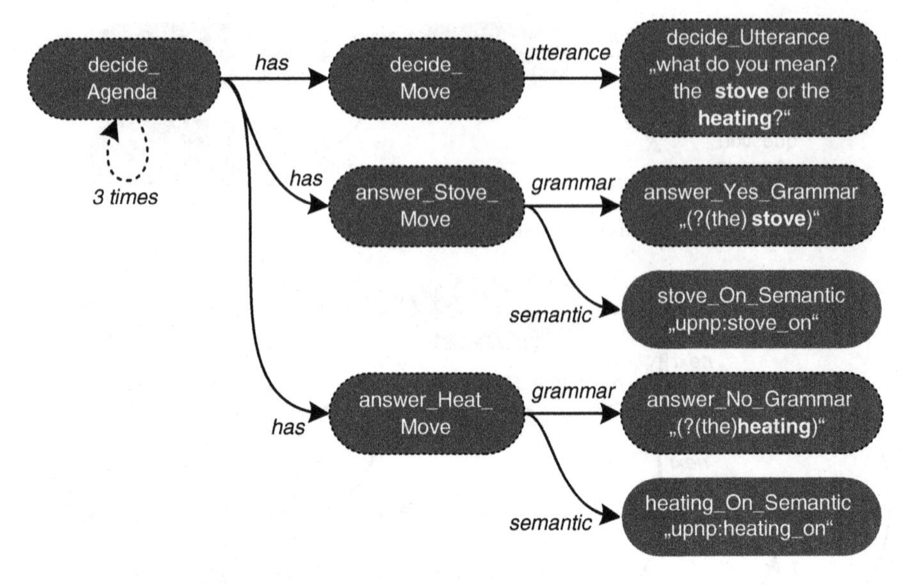

Fig. 4.18 An agenda that realises a decision-making turn. The virtual (i.e. not explicitly modelled) object that are generated on the fly are marked with *dotted lines*

by the ASDM. For generating a new belief, the specific move provides a semantic individual similarly to user and system turns. The question_Agenda points at two subsequent agendas: one that may be processed if the user replies positively and another one if he does not wish to start cooking now. In case of such a negative reply, an agenda that provides a user turn could be activated. The system would listen until the user, for example, utters: "I want to start cooking now!". The two relations *requires* and *mustNot* point at semantics. However, the *requires* relation can only be evaluated as valid if the related semantic individual is referenced by a belief individual, i.e. the meaning of the semantic is shared between user and system.

In contrast to this, *mustNot* only matches if the specified semantic is *not* referenced from a belief individual, i.e. if the meaning of the semantic is not (yet) shared between user and system. An example for such a dialogue situation is the user telling the system to be silent. In that case a specific semantic individual is added to the shared knowledge indicating that only user turns stay activated. All system turns and exchanges would point to the "silence_Semantic" via the *mustNot* relation in order to achieve the desired behaviour. The two clarification turns that have been realised as part of the OwlSpeak system are the decision-making and the confirmation turn. Both turns are realised as exchanges using specific data taken from the respective SDOs.

The main difference to "normal" exchanges is that they are not explicitly modelled as part of an SDO but are automatically generated. Figure 4.18 shows a decision-making turn that looks similar to an exchange as depicted in Fig. 4.17.

We assume that the user utters "it's too hot" during the cooking scenario (see Sect. 3.5). The IE has detected the stove and the heating to be controllable via spoken commands. Both devices provide a grammar that may be used to understand the user input. Thus, the ASDM must generate a clarification turn. Most of the individuals involved in this turn are generated, i.e. they are not predefined as part of an SDO. More precisely, all individuals shown in Fig. 4.18 that are plotted with a dotted line are not predefined.

Instead, a dedicated system-SDO provides dummy individuals that are utilised to generate agendas, moves, grammars, and utterances during runtime. These individuals provide generic information such as "what do you mean" that may be used by the ASDM in combination with, for example, the names of the SDOs that are involved in the conflicting dialogue. Figure 4.18 highlights the human-understandable names of the SDO with bold letters—the user may choose which device he is referring to. Such decision-making turns are repeated three times in case the system receives no useful input. Afterwards, the generated agenda is deleted and the ASDM resumes providing dialogues that correspond to the currently available SDOs. The confirmation turn is generated in a similar way. In case the system is not sure whether it has correctly understood the user input or not, it repeats the most probable input. The user must agree or to deny (cf. Sects. 3.5.2 and 3.6.3).

Besides the fundamental relations and concepts, the class Agenda also allocates several data fields. Similar to the class Move, the Agenda class provides a field *prio* that reflects the urgency of an agenda on a scale from 1 to 100. This priority is used to select the most urgent agenda for generating the upcoming dialogue turn. As mentioned in Sect. 4.2.3.3 there are various possibilities to choose the next dialogue turn. In OwlSpeak we have tested how to utilise the priorities statically (i.e. using the predefined values) and dynamically (i.e. by altering the values during runtime depending on the duration of a dialogue). To be able to manage a dialogue flow, it is not only required to be aware of the subsequent dialogue turns, but also important to know how a dialogue starts and what kind of user input may occur during a dialogue turn (i.e. how spoken commands may be integrated in a conversation). For this purpose the class Agenda provides two expressions:

- The dialogue logic that interprets the model needs an entry point in order to delimitate the search space. The agenda that is used as entry point is marked with the *isMaster* field. This "master agenda" references all other subsequent agendas that are relevant to the actual dialogue flow using the *next* relation. The ASDM checks the requirements of the agendas that are referenced by the *next* relation of the master agenda. Then it decides which agendas must be taken into account to generate an appropriate dialogue turn. The actual spoken dialogue usually starts with the agenda that shows the highest priority. After the master agenda is detected, the decision logic that processes the dialogue model via the *requires* and the *mustNot* relations, is able to select subsequent agendas. In each domain model (i.e. in each SDO) a maximum of one agenda may be defined to be interpreted as master agenda.

- If an agenda should still be available after being processed (i.e. after the user uttered a corresponding statement), the field *respawn* must be set to true. Otherwise, an agenda would not be processed again in order to avoid cycles that prevent the ASDM pursuing the dialogue. In practice, the *respawn* flag indicates that the agenda is used as a spoken command that should always be understood by the system (e.g. "lights on").

A further data field is *variableOperator*. It is used by a built-in scripting language that is able to perform logical operations on semantics and variables. These operations exceed the possibilities of the above-mentioned relations. The scripting approach is an extension to the rigid OWL structures. We further explained this approach in Appendix C. In principle, OwlSpeak is a multi-user system. Thus, all dynamic objects that are generated during an ongoing conversation must be separated user-wise. For this purpose, two sub-classes are part of the SDO: the class BeliefSpace, which is inherited from the class Belief and the class WorkSpace, which is inherited from the class Agenda. Due to the separation of the SDO into static and dynamic knowledge it is sufficient to provide the functionality of storing OWL individuals that are generated during an ongoing dialogue, i.e. that are part of the dynamic knowledge.

During the initiation, the system instantiates an OWL individual for each user in the WorkSpace class and a further OWL individual in the BeliefSpace class. Therefore, we gain a structured user-knowledge scheme. The BeliefSpace is used to sort beliefs and to assign them to users. Each belief therefore is referenced by the *hasBelief* relation from the corresponding beliefSpace individual. Thus, the system separately stores all semantics for each user. Analogously, each workSpace individual references the actual agendas using the *hasAgenda* relation. Furthermore, the workSpaces are utilised as stacks of agendas that need to be processed. If, for example, agenda A is referenced by a *next* relation of agenda B that has already been processed (i.e. the conversational act occurred), agenda A would be a candidate to be chosen as upcoming dialogue turn. Thus, a reference to agenda A is placed in the corresponding workSpace. If agenda A would then be processed successfully, this "to-do" reference would be deleted. Of course, new agendas may also be added to the workSpace during the ongoing dialogue. This dynamic handling of agendas allows for a flexible dialogue management. Since the dynamic part of the domain model alters and evolves during conversation, it is also required to re-initialise the dialogue and in doing so "forgetting" all knowledge that so far has been shared between the user and the system. The classes BeliefSpace and WorkSpace provide a solution to this: In case of such a dialogue reset (for a specific user) the corresponding workSpace individual (together with all references to agendas) and furthermore all beliefs that reference the valid semantics from the user's beliefSpace (together with the beliefSpace itself) are deleted.

Afterwards, two new workSpace and beliefSpace individuals are applied. The referenced agendas of all available master agendas are then added to the newly generated workSpace. Thus, all semantics and variables that have been validated and stored as beliefs are erased. Notably, the variables store the default values that were applied when the SDO has been defined. While the dialogue progresses, agendas will be deleted from the workSpace, new agendas will be applied, and semantics and/or variables will be linked as new beliefs in order to be part of the user's beliefSpace. The class History provides a dynamic log that records which agenda has been marked as to-do, the priority of the agenda, and a timestamp defining the actual time it has been processed. The relation *forAgenda* provides a pointer to the respective agenda and the three data fields addTime, agendaPrio, and procTime to store the log data described above. A second relation points at the actual workSpace of the user in order to provide a user-wise log history.

The combination of the classes and relations described in this section compose a powerful logical framework that may be used to define spoken dialogue domains, which are meant to be combined with other dialogue domains during runtime. Thus, the framework facilitates the realisation of the environment-related aspect of adaptation. The approach allows multitasking since it is possible to continuously combine different (dialogue) tasks that may occur in parallel. Of course, the framework also realises aspects of user-centred adaptation: the functionality of dialogue control on part of the user is a unique characteristic of our system. For example, with a rule-based dialogue manager such as Ravenclaw (Bohus and Rudnicky 2009) it would be difficult to realise similar functionality. Besides the benefits of the model-based approach to ASDM we also encountered drawbacks regarding the inflexible structure of OWL. Hence, in Appendix C we present an extension of the OWL-based SDO. The proposed scripting language facilitates the modelling of dialogues, while staying fully compatible with the OWL standard. In the following section we illustrate how a dialogue domain may be modelled and how it can be combined with other dialogue domain models using the proposed framework. As an example we refer to the initial part of the conversation presented in Sect. 3.5.

4.3.3 An Elaborated Example

The framework defined in the previous section is on one hand powerful when it comes to express interdependent tasks and on the other hand straightforward when it comes to develop spoken dialogues. In this section we demonstrate how the framework may be utilised to implement a dialogue that could be used to perform the "Prepare dinner" task as described in Table 3.5. For main applications interaction regarding information retrieval is crucial. For this purpose, question–answer dialogues are perfectly suited to be expressed using the OwlSpeak ASDM framework.

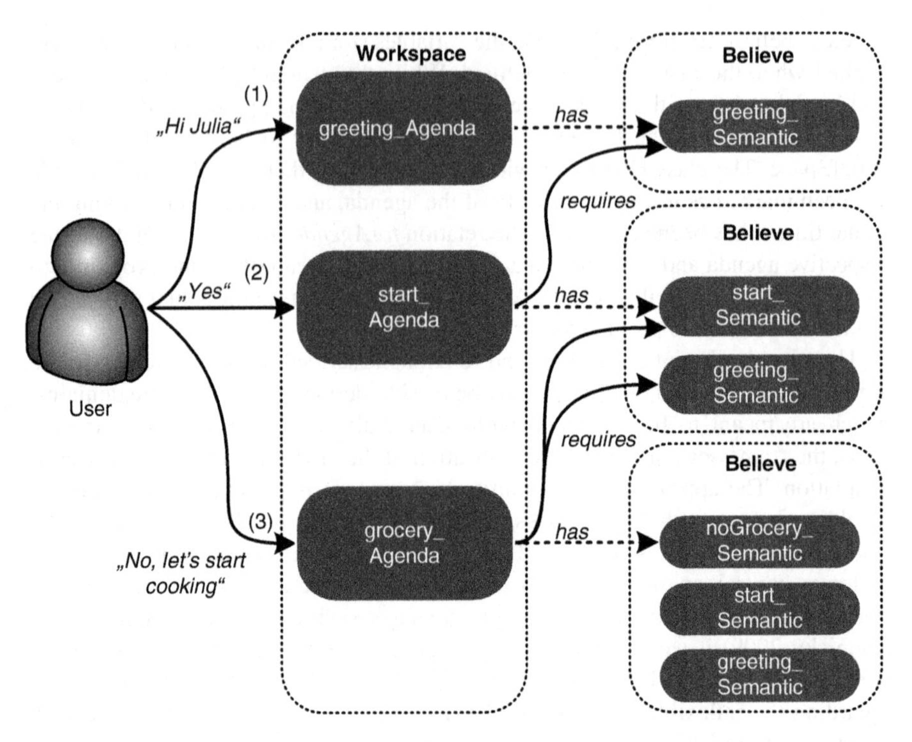

Fig. 4.19 The agendas as part of the workspace and their relationships to the believes

4.3.3.1 A Basic Dialogue

In the following we present three agendas, each representing a question by the system and possible answers by the user. To keep the example as simple as possible we leave the (spoken dialogue) tasks that may also be part of the IE, such as command dialogues for controlling the environment, aside. How can the SDO be utilised to express the required dialogue? Figure 4.19 presents the three agendas that are necessary to implement the exemplary dialogue situation that comprises of the greeting and the introduction of the "Prepare dinner" dialogue. Furthermore, the figure illustrates how the agendas are "sorted" in order to achieve a suitable dialogue flow as described in the scenario. Each of the agendas provides specific knowledge encoded as a semantic value. Once an agenda is being processed, e.g. the user positively answered the first question, the ASDM adds the start_Semantic to the appropriate believe.

The subsequent agenda, in this case the grocery_Agenda, requires this semantic value as part of the system's believe space. The three agendas express the following dialogue-related information:

Greeting_Agenda This agenda allows the system to understand a greeting uttered by the user. It is realised by a user turn and therefore the system passively waits for input. The activated grammars may, for example, match "hi Julia". As long as the user does not greet the system, the dialogue does not progress. Once the user greets the system, a newly applied believe individual points at the greeting_Semantic individual. It is then marked as shared knowledge.

Start_Agenda This agenda requires the greeting_Semantic to be provided by the greeting_Agenda. The ASDM asks the user whether he wishes to start the dinner preparation task. This question may be answered with "yes" or "no". If the user positively replies, the system proceeds with the next agenda, in this example the grocery_Agenda. If the user denies, the ASDM switches to a passive agenda that waits for the user uttering, for example, "start dinner preparation now". Such an agenda is realised in a similar way as the "Lighting control" task presented in the following section and is not further discussed here.

Grocery_Agenda The system is aware of two facts: (1) the user initiated the dialogue (i.e. the greeting_Semantic is part of the beliefs) and (2) he wishes to start preparing the dinner (i.e. the start_Semantic is part of the beliefs). For this reason, the grocery_Agenda may be used to generate the next dialogue step. In order to find out, which recipe could be selected for preparing the dinner, the system wants to know if the user plans to go to the grocery. In our example the user does not want to go there but rather wants to start cooking.

The process of detecting agendas that are suitable for generating a new dialogue step is carried out in a cyclic manner. Theoretically, the OwlSpeak ASDM is never deactivated. However, two reasons for entering a state of inactivity exist: either there is no further agenda available to be carried out or no agenda can be selected since not all required semantics are referenced from the beliefs. This approach is very flexible because one or more than one agenda may be active at one particular point in time depending on the knowledge that is shared between user and system. After all semantics that are required to, for example, select a recipe, are collected, the ASDM is able to pass the knowledge to a subsequent dialogue that is related to the initial one. If external devices or services should also take advantage of the collected knowledge, a specific format and protocol must be defined [e.g. UPnP (ISO 2008)]. The prototype supports UPnP, but several other formats are also feasible to exchange data with external services or devices. To summarise the fundamental mechanisms compounding the OwlSpeak ASDM, Fig. 4.20 depicts the main process of receiving an utterance on part of the user, updating the underlying SDO(s) and generating a new dialogue view (cf. Sect. 4.2.2—the figure corresponds with the "work" operation).

The main process begins with either a grammar match or a non-understanding (i.e. a nomatch). The figure shows the optional dialogue management techniques in small font (clarifications and adaptive recognition). If the grammar detects a valid user input, the corresponding semantic individual is added to the beliefs stored in the respective SDO. If the grammar-providing agenda points to any subsequent agendas, the ASDM adds these agendas to the workspace of the user, i.e. to his "to-do" list. Afterwards, the agenda itself is being dropped from the workspace. At this point, agendas used to understand commands that are marked as "respawns" are not

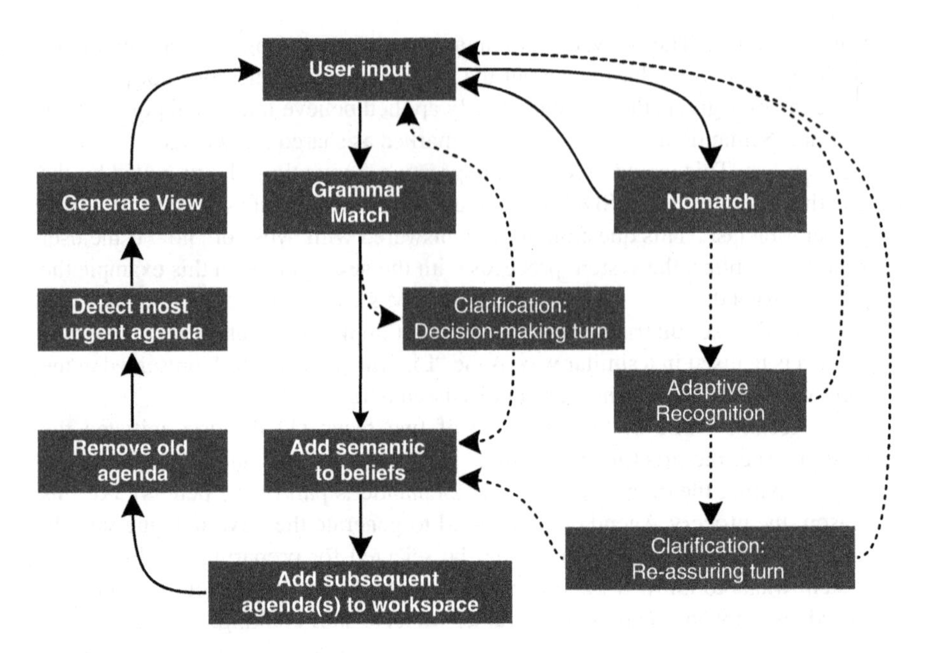

Fig. 4.20 The main process of understanding the user input, updating the knowledgebase, and generating a new dialogue view

dropped since the system must still understand them during the subsequent dialogue turn. Then the most urgent agenda is selected for generating the next dialogue view. At this point, things have come full circle and the system is ready to receive the next user input.

In case of a nomatch, the adaptive recognition methods that have been introduced in Sect. 4.2.3.4 may be applied. If a meaningful user input is being detected, a clarification turn is initialised and, if the user agrees, the semantic value is added to the beliefs. All arrows that lead back to the user input in Fig. 4.20 indicate a request command as defined in Sect. 4.2.2. Thus, basing on the actual SDOs, an updated dialogue view is generated. A second type of non-standard system behaviour is caused by ambiguities. If the actual grammar is ambiguous, a decision-making turn is being applied to the normal dialogue management process. If the user is able to clarify the actual dialogue turn, the semantic is added to the beliefs as it would have been done without the occurrence of an ambiguity. Before we present some examples of dialogue views in Sect. 4.4, we provide a closer look at the capacities of the framework in the following section. We focus on establishing logical dependencies between individuals and on using variables to either incorporate external knowledge into grammars and utterances or to expose collected knowledge that may be used by external entities.

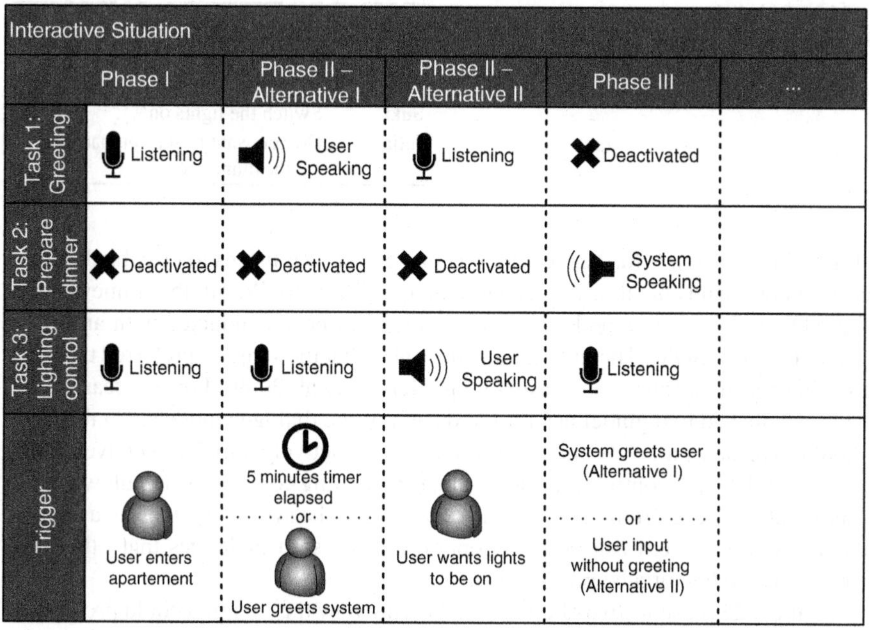

Interactive Situation	Phase I	Phase II – Alternative I	Phase II – Alternative II	Phase III	...
Task 1: Greeting	Listening	User Speaking	Listening	Deactivated	
Task 2: Prepare dinner	Deactivated	Deactivated	Deactivated	System Speaking	
Task 3: Lighting control	Listening	Listening	User Speaking	Listening	
Trigger	User enters apartement	5 minutes timer elapsed or User greets system	User wants lights to be on	System greets user (Alternative I) or User input without greeting (Alternative II)	

Fig. 4.21 An interactive situation that may occur with two alternatives

4.3.3.2 An Interdependent Dialogue

In this section we discuss how the framework presented in Sect. 4.3 may be utilised to define a set of spoken dialogue domains that can be used interdependently within an IE. We assume that a typical situation in a fictive world where a user lives together with his IE is when he arrives home. An example of such a situation is described as part of the dialogue scenario in Sect. 3.5. However, several other tasks besides the "greeting" task that lead to the dinner preparation dialogue may run in parallel. Each task provides the possibility of spoken interaction. Since one of the main duties, an IE has to handle, is to control tasks (i.e. the system should provide possibilities to facilitate the user's access to various functionalities), it is necessary to provide a probably varying set of spoken commands that the system may interpret and execute. An example of this behaviour is the user telling the system to switch the lights on after entering the apartment.

Figure 4.21 shows a set of three interactive tasks shaping an interactive situation as it may occur at the beginning of the scenario described in Sect. 3.5. Since the ASDM adapts to the context, it must receive triggers from the surrounding IE to change its state. Therefore, the initial phase is triggered by an "user enters apartment" event. This event may happen only once a day and/or when the user has left the apartment for a specified period depending on the configuration of the IE. In our example, the ASDM must wait until the user greets the system

Table 4.6 A dialogue snipped that may occur during the proposed interactive situation

Speaker	Utterance	Task
Suki	Hello Julia!	1
Julia	Hi Suki!	1
Suki	Switch the lights on!	3
Julia	Do you want to start preparing the dinner now?	2

(Fig. 4.21(Task 1)). It further activates a control task that listens to possible lighting control commands the user may utter (Fig. 4.21(Task 3)). Within the framework of the ATRACO project, a preliminary evaluation has been conducted with an initial version of the ASDM. The investigation revealed that the subjects preferred the SDS to behave as unobtrusive as possible (van Helvert et al. 2009). For this reason, we have established two guidelines for the design of the dialogue models: (1) the SDS should to behave passively and (2) it should not proactively initialise a conversion if this is avoidable. A control task such as Task 3 (lighting control) by default waits for user input. However, if the user initialises talking to the system by uttering a spoken command, the system can take this opportunity to start dialogues that otherwise would have to be initiated proactively.

Figure 4.21 presents two alternatives showing how the situation could proceed in Phase II. Alternative I contains two triggers that may allow the system to perform Task 1: the 5-min-timer elapsed since the user entered the apartment or—probably the more usual case—the user greets the system. As mentioned above, the reason for such a 5-min-timer is that the system should act as unobtrusively as possible. Note that Task 3 is still active since the system handles more than one interactive task in parallel. If one of the two triggers is actuated, the system greets the user and adds a semantic value (greeting_Semantic) to the knowledgebase (technically the system may also use a variable for the same purpose). This would allow Phase III to start. Table 4.6 shows a possible conversation that may occur utilising the proposed set of SDOs for the three tasks. Alternatively, Phase II could pass off conditioned by the user telling the system to switch the lights on. This would make Task 1 obsolete—the system should not greet the user in response to a spoken command. It would be more natural if the system skips the greeting task and instead activates the proactive Task 2 "Prepare dinner".

Figure 4.21 shows Phase III constituted by the additionally activated Task 2 and the still running Task 3. The preceding greeting task has either become obsolete or has already being processed. At any time *t* user or the IE may dynamically activate or deactivate all spoken dialogue tasks, the ASDM is able to perform. In the following, we present the dialogue models and their interplay for Task 1, 2, and 3 that form the described interactive situation. In this context it is necessary to explain the role of variables within the ASDM framework. All causal dependencies between the different tasks that should be available and modifiable from outside (i.e. from the IE) are handled by variables and their corresponding logical operations. Figure 4.22 presents the individuals that are necessary to implement Task 1, the adaptive greeting.

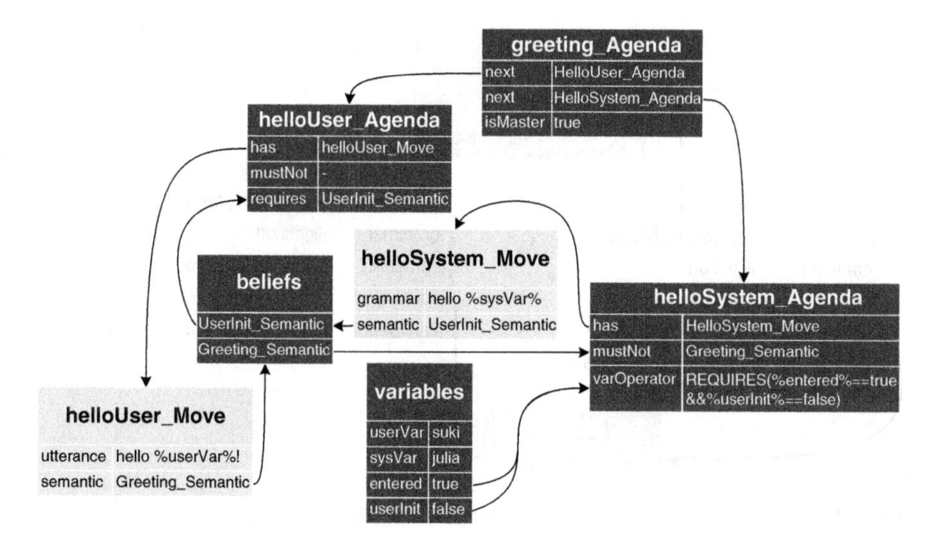

Fig. 4.22 The individuals of the adaptive greeting task and their interplay. The static individuals involved are *light grey dyed*

The figure shows the master agenda (greeting_Agenda) on top. It provides two related follow-up agendas, both linked by the *next* relation. While the ASDM interprets these relations, it adds the two individuals helloSystem_Agenda and helloUser_Agenda to the actual to-do list (i.e. the specific workspace). If the requirements of, at least, one of these agendas are fulfilled, a dialogue snippet with the according grammar and/or utterance can be generated. In the initial dialogue state both agendas cannot be used for dialogue generation since the "REQUIRES" function defined in the *varOperator* field of the helloSystem_Agenda cannot be validated as true—the user must enter the environment and/or utter a command before. Figure 4.22 shows a state of the SDO when the greeting dialogue has already been carried out: as described in the exemplary interactive situation (Fig. 4.21) the greeting task should not start before the user enters the apartment. The related variable "entered" must be set to true if this happens. For the next phase two alternatives may occur. The most usual way the situation could evolve would be that the user greets the system directly after entering the apartment. To be able to react on such a user initiated behaviour the IE must provide the information that the user has entered the apartment. This would update the ASDM variable "entered". The helloSystem_Agenda is now accessible and the system listens to the user, who may utter "hello Julia".

This constellation comes to complete Alternative I of Phase II by adding the userInit_Semantic to the beliefs. This allows the ASDM to utilise the helloUser_Agenda to generate the next dialogue turn. As a result, the system utters "hello Suki". Furthermore, if the user misses to greet the system, for example, a timer-service may be utilised to indicate when the system should greet the user. This timer is part

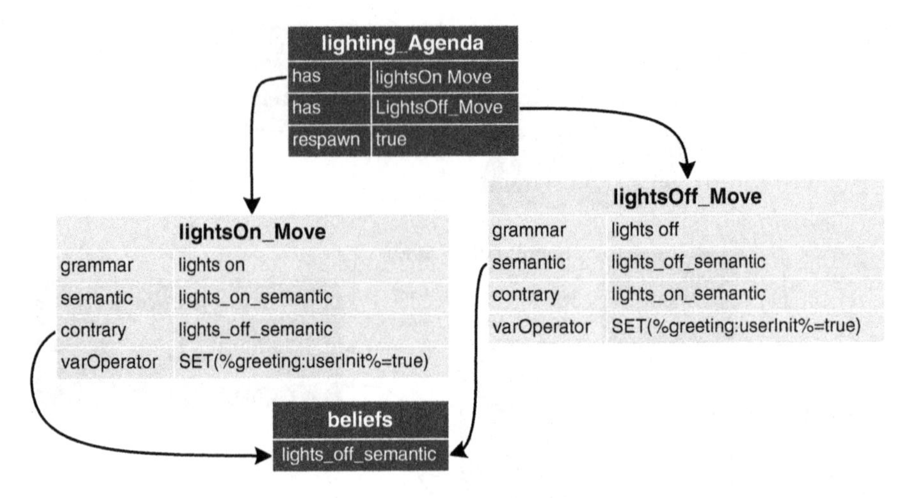

Fig. 4.23 The individuals of the adaptive lighting task and their interplay. The static individuals involved are *light grey dyed*

of the IE and, depending of the configuration, may be activated when the user enters the apartment. We promise ourselves that such behaviour is less obtrusive for the user. Other triggers aside from a timer could also be utilised. Conceivable are also sensors that are attached to the user's coat indicating when the user takes it off. Also specific areas where the system starts communicating with the user, for example, after entering the living room may be realised with location assessing systems such as Ubisense (2011).

In Fig. 4.21 a differing course of the interactive situation is described (Alternative II). The user does not greet the system but instead switches on the lights via speech. This makes a system initiated greeting obsolete and obviously it is not necessary any more to wait for the user greeting the system. Thus, Task 3, which is always running passively and listens to appropriate user commands, triggers the ASDM to deactivate Task 1 and to activate the next task that is within the pipeline of the ASDM, the dinner preparation task. Such passive and persistent tasks must be handled separately. As shown in Fig. 4.23 they are marked as *respawn* and are running in parallel with a normal interactive task. The inter-task communication that enables the ASDM to activate/deactivate tasks is realised by the concept of variables: both the lightsOn_Move and the lightsOff_Move encode a *variableOperator* field. This field is used to modify the variables that affect the dialogue flow of the greeting task (via the SET operator). This triggers the system to enter Phase III, where the "Prepare dinner" task can be executed.

The lighting_Agenda can still be activated as part of this phase; however, the system would not greet the user since it either already happened or it has become obsolete. Our scenario describes several dialogue flows that may be influenced by the variables concept. We have chosen to use variables in this context since they can be modified from external entities that are part of the IE. Contrary to this, the semantics concept as used in Sect. 4.3 is meant to be used for the encoding of internal (i.e. dialogue inherent) knowledge. Therefore, the variables are a more complex and at the same time a more flexible concept. They may be used to activate new tasks or to deactivate obsolete ones either by logical operations on specific variables as described in the previous paragraphs or by UPnP commands as described in Sect. 4.2.1. Since the entire OwlSpeak ASDM framework is UPnP-wrapped, all administrative operations that are necessary to establish an interactive situation as described above, may also be triggered by external IE members.

In this section we have presented the domain model, which forms the main source of knowledge that highly influences the generation of the dialogues. In Sect. 4.2 we have presented the inherent logic of the ASDM, which works on the domain model and decides how dialogues are kept coherent, how they may be combined, and how they can be carried out in general. The dialogue itself bases on the fundament that is compound by the dialogue models and the decision logic. As previously mentioned, OwlSpeak uses VoiceXML to express the dialogue snippets that are periodically updated depending on the actual state of the domain model. In the following section we present the view layer that allows to generate the dialogue snippets, which are finally interpreted by the SDS.

4.4 The View

The modular and flexible OwlSpeak ASDM framework results from the rigid distinction between the decision logic, the domain knowledge, and the interface the user perceives. In this section we present this interface that is realised as part of the MVP pattern—as the view. Due to the decoupled system architecture OwlSpeak is able to generate a variety of dialogue describing formats. For our prototype we implemented a VoiceXML-based view that is discussed in the following section. In Sect. 4.4.3 we provide some exemplary alternatives that may also be useful for specific tasks and for the integration into specific SDSs. The underlying mechanisms of OwlSpeak generate the dialogue definition the SDS is able to interpret during runtime in a cyclic manner. We propose the use of so-called dialogue snippets that represent the dialogue turns a conversation may consist of. In the following we present the VoiceXML descriptions of the turns that are utilised to generate adaptive spoken dialogues for IEs.

4.4.1 Basic Turns

As described in Sect. 3.5.2 we use four fundamental types of turns for the dialogue generation: user turns, system turns, exchanges, and clarifications. Depending on the actual state of the SDO(s), the presenter selects an appropriate template and passes it to the SDS, which is responsible for recognising the user input and for synthesising the appropriate output. As depicted in Listing 4.1 a user turn consists of a grammar defining the possible input of the user combined with a property value providing a timeout. This timeout is used to define the period, the system waits for input uttered by the user.

If no input is provided, a specific event referred to as "noinput" is thrown and the ASDM is requested to provide a newly generated dialogue snipped that correlates with the current dialogue state. In case the dialogue state encoded by the SDOs has not been modified, a similar dialogue snipped is generated again. Therefore, it is possible that the ASDM behaves passively until the user initiates the interaction. If the user utters a command that is not covered by the activated grammar, a different event is thrown: a "nomatch". This event may be handled differently by the ASDM. The user utterance may be analysed by detecting keywords or by involving a semantic knowledgebase (see Sect. 3.6.3). If the user utters a command that is covered by the grammar—in Listing 4.1 "hi Julia"—the actual input is passed to the ASDM that updates the SDO accordingly and probably generates a new dialogue snipped. A VoiceXML snippet that realises a user turn is also used for realising a "standby" state. It allows to define a "magic word" for waking the system up. Otherwise, only the environment would be able to re-initiate the voice interface.

```
1  <vxml version="2.1">
2   <property name="timeout" value="5s"/>
3   <form>
4    <var name="com"/>
5    <var name="agenda" expr="'userSaysHello_agenda'"/>
6    <field name="move">
7     <grammar type="application/x-gsl">
8      [[hi julia]]{<move "userSaysHello_move">}]
9     </grammar>
10    <catch event="nomatch noinput">
11     <assign name="com" expr="'request'"/>
12     <submit next="http://server.de/owlSpeak/owlSpeak" namelist="
           com agenda"/>
13    </catch>
14    <filled>
15     <assign name="com" expr="'work'"/>
16     <submit next="http://server.de/owlSpeak/owlSpeak" namelist="
           com move agenda"/>
17    </filled>
18   </field>
19  </form>
20 </vxml>
```
Listing 4.1 A user turn dialogue snipped in VoiceXML

Listing 4.2 shows a dialogue snipped that may be generated to appropriately react on the user input that occurred in Listing 4.1: the system greets the user. This type of dialogue (a system turn) does not need a specific timeout or any other trigger to communicate with the presenter layer of the ASDM. After the system utters a message, it assumes that the user has correctly understood and a new dialogue is requested. However, the model-based approach to ASDM in general is suited to incorporate a "repeat" command enabling the user to hear the last system output again.

```
1  <vxml version="2.1">
2   <form>
3    <var name="com"/>
4    <var name="agenda" expr="'sysSaysHello_agenda'"/>
5    <block>
6     <prompt bargein="false">
7      hi suki
8     <break/>
9     </prompt>
10    <assign name="com" expr="'work'"/>
11    <submit next="http://server.de/owlSpeak/owlSpeak" namelist="com
          agenda"/>
12   </block>
13  </form>
14 </vxml>
```

Listing 4.2 A system turn dialogue snipped in VoiceXML

The class History that has been presented in Sect. 4.3 may be used to spot the last agenda that has been involved in dialogue generation. This agenda can then be utilised to generate the last dialogue snippet again. However, this functionality is not realised as part of the prototype yet. After the greeting is finished, an exchange as shown in Listing 4.3 can be provided by the ASDM. Here the system combines a system turn and a user turn in the form of a question–answering adjacency pair. Similar to a user turn, the system waits a couple of seconds until the ASDM is requested to send a new dialogue snipped depending on the current status. If the state of the SDOs has not been modified, the exchange is generated again. As a result, the system asks the question again. In practice, the question would be repeated until either the system receives an input or the knowledgebase is modified and the exchange is not required any more. Analogously to a user turn, an exchange may be enhanced with the adaptive recognition methods presented in Sect. 3.6.3.

```
1  <vxml version="2.1">
2   <property name="timeout" value="5s"/>
3   <form>
4    <var name="com"/>
5    <var name="agenda" expr="'prepareDinner_agenda'"/>
6    <field name="move">
7     <grammar type="application/x-gsl">
8      [[[yes]]{<move "dinnerYes_move">} [[no]]{<move "dinnerNo_move"
          >}]
9     </grammar>
10    <prompt bargein="false">
11     Do you want to start the dinner preparation now?
12    <break/>
13    </prompt>
14    <catch event="nomatch noinput">
15     <assign name="com" expr="'request'"/>
16     <submit next="http://server.de/owlSpeak/owlSpeak" namelist="
          com agenda"/>
17    </catch>
18    <filled>
19     <assign name="com" expr="'work'"/>
20     <submit next="http://server.de/owlSpeak/owlSpeak" namelist="
          com move agenda"/>
21    </filled>
22   </field>
23  </form>
24 </vxml>
```

Listing 4.3 An exchange dialogue snipped in VoiceXML

Listing 4.4 shows a modified "nomatch tag" that matches all user input that cannot be covered by the grammar. The recorded input which is provided by the VoiceXML built-in variable "lastresult$.recording" is transferred to the ASDM via a POST submit. Here, the input can be analysed again. In case any of the methods proposed in Sect. 4.2.3.4 detects a dialogue domain and a corresponding move that matches the utterance, a confirmation turn (see following section) is generated automatically. This confirmation is required in order to avoid misunderstandings.

```
1  ...
2  <nomatch>
3    <assign name="nomatchRec" expr="lastresult$.recording"/>
4    <assign name="event" expr="_event"/>
5    <assign name="com" expr="'request'"/>
6    <submit namelist="com user counter event nomatchRec" enctype="
         multipart/form-data" method="POST" expr="'http://server.de/
         owlSpeak?com='+com+'&event='+event"/>
7  </nomatch>
8  ...
```

Listing 4.4 The modified nomatch tag allowing the system to perform an additional analysis of the user input

The optimal SDS behaviour is to understand the user correctly. In that case the user input is covered by the grammar. Thus, it is not necessary to transfer the recorded utterance to the ASDM. Only the information that is needed to update the SDO is submitted in this case: the unique identifier of the SDO and the move that provides the actual grammar. Taking this information into account, the ASDM adds the according semantic value(s) to the beliefs and triggers the generation of a new dialogue turn. This fundamental system functionality is outlined in Sect. 4.2.2. In the next section we show how the clarification turns are generated automatically.

4.4.2 Clarification Turns

Two main types of clarification turns have been introduced in Sect. 3.5.2: decision-making and confirmation turns. The former is required in case the ASDM detects an overlapping grammar, i.e. a user input that may be recognised by taking more than one move (with a specific grammar) into account. In Sect. 4.2.3.2 we describe the ability of the ASDM to automatically generate a decision-making turn for conflict resolution. In order to generate such dialogues, the SDOs have to encode unique human-understandable identifiers. Uniqueness in this sense also includes the avoidance of homonyms. These identifiers are utilised to generate both the names provided as part of the question (i.e. the prompt) and the grammars that are necessary to understand the user's response (cf. Listing 4.5). The template that is used to generate Listing 4.5 is customised for conflicts that may occur in command-and-control dialogues. Besides the uniqueness of the dialogue domain identifier, some constraints must be taken into account here: the ASDM should not list too much alternatives since the user cannot keep an unlimited number of devices or services in mind. We do not assume that this is a general limitation of our prototype. Usually, an IE (within the context of the user) does not consist of more than five similar devices (e.g. lights) that may be involved in the conflict.

```
1  <vxml version="2.1">
2   <property name="timeout" value="5s"/>
3   <form>
4    <var name="com"/>
5    <var name="agenda" expr="'question_agenda'"/>
6    <field name="move">
7     <grammar type="application/x-gsl">
8      [[(table lamp)]{<move "tableLamp_on_move">}[(ceiling light)]{<
         move "ceilingLight_on_move">}]
9     </grammar>
10    <prompt bargein="false">
11     Which one?  Table lamp or ceiling light?
12    <break/>
13    </prompt>
14    <catch count="3" event="nomatch noinput">
15     <assign name="com" expr="'request'"/>
16     <submit next="http://url/owlSpeak" namelist="com agenda"/>
17    </catch>
18    <filled>
19     <assign name="com" expr="'work'"/>
20     <submit next="http://url/owlSpeak" namelist="com move agenda"/
         >
21    </filled>
22   </field>
23  </form>
24 </vxml>
```
Listing 4.5 An automatically generated decision-making turn

Furthermore, the identifiers should be easily repeatable by the user, i.e. they should not be too complex (e.g. "light number 73425 below the table" would not be a good choice). In order to avoid an unlimited loop in case the system does not recognise the user's input correctly, the system asks the user three times to solve the conflict. If this attempt is not successful the system does not carry out any command and resumes a probably existing previous dialogue. Of course, this kind of conflict resolution may also be used to solve conflicts that occur as part of a conversation. However, the automatically generated question would not sound natural in such a situation. In case the user and system discuss, for example, the modification of a recipe, the system would perhaps ask: "Which one, the pizza or the fish". To generate prompts (and according grammars) that sound natural in all contexts is not a trivial task and would require further information provided by the SDOs.

A second type of clarification turns are confirmation turns. These are generated by the ASDM in case the adaptive understanding mechanism presented in Sect. 4.2.3.4 provide a positive result (under a specific probability). Listing 4.6 shows such a turn. Here, at the first attempt, the ASDM tries to map the user input to an utterance that is part of the dialogue domain. This utterance is then used to generate a question such as "have you said 'lights on'". The user replies positively or denies. In case he denies the ASDM may generate a similar question regarding an input that is less probable (cf. Line 35–37 in Listing 4.6). In our example a subsequent question could be "have you said 'telly on'". If there is no further input (that is above a specific threshold of probability) detected, the ASDM ignores the input and resumes with the dialogue as currently defined by the states of the SDO(s).

```
1  <vxml version="2.1">
2   <property name="timeout" value="5s"/>
3   <form><var name="com"/>
4    <var name="move" expr="'light_a1n1d_LightsOn_move'"/>
5    <field name="LightsOn_move">
6      <grammar type="application/x-gsl" xml:lang="de-de" mode="voice
           ">
7        [  [yes yep correct]  { <LightsOn_move "true"> }
8               [no wrong nope]  { <LightsOn_move "false"> }   ]
9      </grammar>
10      <prompt bargein="false">have you said 'lights on'?
11     </prompt>
12     <catch event="nomatch noinput">
13       <prompt bargein="false">
14         pardon?
15       </prompt>
16     </catch>
17     <catch count="3" event="nomatch noinput">
18       <prompt bargein="false">
19         sorry, I din't get that.
20       </prompt>
21       <assign name="com" expr="'request'"/>
22       <submit namelist="com agenda user" next="http://url/owlSpeak
             "/>
23     </catch>
24       <filled>
25         <if cond="LightsOn_move == 'true'">
26           <assign name="com" expr="'work'"/>
27           <assign name="speak" expr="LightsOn_move$.utterance"/>
28           <submit namelist="com move agenda speak user" next="http
               ://url/owlSpeak"/>
29           <else/><goto next="#TvOn_move"/></if>
30       </filled>
31     </field>
32         <field name="TvOn_move">
33         ...
34         </field>
35    </form>
36 </vxml>
```

Listing 4.6 An automatically generated confirmation turn

Our framework allows for a variety of special purpose turns besides the ones we introduced in this section. Conceivable are turns that realise a "repeat" command, a "help" command, or a generic "skip this" utterance. The open design of the ASDM also allows for different formats that are necessary to combine the ASDM with further SDSs. In the following section we present some suggestions for other useful formats that may also be used as output of the view layer.

4.4.3 Alternative Views

Thanks to the modular framework of the MVP pattern it is straightforward to define alternative views. For our prototype we decided to generate the W3C standardised VoiceXML in combination with the proprietary Nuance GSL grammar (Nuance 2008). To stay compatible with other dialogue description languages such as SALT [Speech Application Language Tags (Wang 2002)], the dialogue generating logic is completely separated from the view. The ASDM also supports several other grammars that are usually defined inline as part of the dialogue description. Besides Nuance GSL, the ASDM supports formats such as the Speech Recognition Grammar Specification (SRGS, Brown et al. 2004) or the JSpeech Grammar Format (JSGF, Hunt 2000). For more advanced approaches it would also be possible to define a message-oriented recognition and synthesis framework in order to combine the ASDM with an agent-based SDS, for example, such as Olympus (Bohus et al. 2007) is.

```
1  <HTML>
2    <Body>
3      The system says:<br/>
4      Utterance:Do you want to start preparing the dinner?<br/>
5      The system understands:<br/>
6      grammar:
7      <a href="./owlSpeak?com=work&agenda=question_agenda&move=
           yesnoquestion_a1n1d_no_move">(no)</a><br/>
8      grammar:
9      <a href="./owlSpeak?com=work&agenda=question_agenda&move=
           yesnoquestion_a1n1d_exit_move">[leave me alone]</a><br/>
10     grammar:
11    <a href="./owlSpeak?com=work&agenda=question_agenda&move=
          yesnoquestion_a1n1d_yes_move">(yes)</a><br/>
12     grammar:
13     <a href="./owlSpeak?com=work&agenda=question_agenda&move=
           light_a1n1d_on_move">(lights on)</a><br/>
14     grammar:
15     <a href="./owlSpeak?com=work&agenda=question_agenda&move=
           light_a1n1d_off_move">(lights off)</a><br/>
16     ...
17   </Body>
18 </HTML>
```

Listing 4.7 The view in non-verbal "browsing" mode using HTML

Building upon the structure and the underlying ideas of OwlSpeak, it is also possible to extend the system to a multimodal ASDM. It is possible to generate, for example, documents by using XHTML+Voice (X+V, see Axelsson et al. 2001) or EMMA (Johnston et al. 2009) instead of VoiceXML. For this purpose, the knowledgebase (i.e. the basic definition of the SDO) must be extended in order to allow for storing the information that must be provided using text or graphics and the input that must be accepted using gestures, written statements, or hypertext methods. However, our current prototype can only interact via voice *or* by using a hypertext browsing mode. Listing 4.7 shows the hypertext view that is written in HTML. The utterances are written as text instead being synthesised by the TTS and the grammars that may be used to define the recognisable input are rendered as hypertext links. This kind of view is perfectly suited for testing and rapid prototyping of spoken dialogues.

4.5 Conclusion

In this chapter we explained the conceptual basis of the OwlSpeak ASDM. The selection of the MVP architectural pattern was driven by the intention to develop a flexible and adaptive dialogue management framework. The topmost layer, the presenter, is discussed in Sect. 4.2. The implemented functionalities follow-up the concepts we have presented in Chap. 3. We summarise these functionalities as follows:

- The provision of a two-sided interface is the main purpose of the presenter layer. One side provides the interface between the view and the presenter. The other side provides the interface to the IE and to the user who may wish to manually control the dialogue management.
- The basic functionality of the presenter is to generate spoken dialogues consisting of dialogue turns or conversational acts that should subsequently be carried out. These dialogues are generated in a cyclic way so as to guarantee for an always updated view (i.e. dialogue description).
- Apart from the fundamental functionalities that are required for a working prototype, we implemented a set of enhanced methods for SDM within the IE context:

  Concurrent Dialogues The ASDM is able to activate several dialogues in parallel in order to provide speech interfaces to different devices and services populating an IE. Since this population varies during the evolution of the IE, the ASDM must be able to dynamically adapt its set of dialogue models to the actual demand.

  Conflicting Dialogues Since the system allows the provision of more than one dialogue domain in parallel, it is also mandatory to support dialogue domains to overlap. Here, the OwlSpeak ASDM is able to automatically detect ambiguities and to generate dialogues for conflict resolution.

  Prioritisation of Dialogues To enable the ASDM to generate both concurrent and conflicting dialogues, prioritisation methods are required. To allow for a multitasking spoken dialogue interface the ASDM is able to prioritise dialogues of different domains. For the prototype we implemented a dynamic prioritisation approach. Its purpose is to avoid dialogue interruptions as best as possible. Since the urgency of suppressed dialogues may grow over time, a specific threshold must be met before an ongoing dialogue is interrupted.

  Adaptive Understanding To improve multitasking of spoken dialogues our work also focusses on techniques for adaptive understanding. The main reasons for applying these techniques are the requirements arising from the dynamic combination of different domains. Therefore, the ASDM is able to detect the most suitable dialogue domain and to automatically derive the meaning of the user input.

The second layer, the ASDM consists of, is the domain model. It is presented and discussed in Sect. 4.3. One of its most important characteristics is its capability to express both the structure and the actual state of a spoken dialogue. The structure of a dialogue is expressed using a set of classes for utterances (statements by the system), grammars (statements by the user), and semantics (the logical meaning of a dialogue step).

The classes may be combined with each other and reused within different domains. The state of a specific dialogue is persistently stored using an Information State-based approach. The information that is shared between user and system (i.e. both user and system are aware of the information) is stored as part of the Belief class. Here, the semantic meanings are used to express the shared information. Subsequent dialogue acts may depend on specific semantics that must be contained within the Belief. In Sect. 4.3.3 we discuss how dialogue models can be designed using, for example, the standard OWL-modelling framework Protégé. A basic SDO dialogue definition is presented to show how the fundamental modelling guidelines work. In a second, more complex example we explain how different dialogues may depend on each other if this is necessary to realise specific dialogue situations.

The third layer of the ASDM is the view. It is used to provide the dialogue the ASDM generated before. Here, the flexibility of the MVP pattern leads to a broad range of formats, supported by the ASDM [e.g. VoiceXMl (Oshry et al. 2007), SALT (Wang 2002), GSL (Nuance 2008), JSGF (Hunt 2000), XHTML+Voice (Axelsson et al. 2001)]. The spoken dialogues we designed for the evaluation sessions are expressed using VoiceXML (see Appendix A). Additionally to voice-based dialogues, it is also possible to generate non-verbal views. To allow for efficient testing of the OwlSpeak ASDM we developed a HTML-based view that can be used without an SDS. An approach for future development may lead to the integration of EMMA (Johnston et al. 2009) into OwlSpeak. To allow for a multimodal interface not only the view would have to be changed: the domain model may also be extended to support GUI specific information (e.g. the colour of button). With respect to the seven levels of adaptation that have been taken into account for the implementation of the prototype, we have realised a working framework for managing spoken dialogues within IEs. The OwlSpeak ASDM prototype together with all resources is released on Soureforge under the General public License (GPL) to be available to the broader scientific community.[1]

After the technical system foundation presented in this chapter, we discuss the evaluation sessions that have been conducted with OwlSpeak in the following chapter. We start with a report on the system evaluation that has been carried out before the ASDM was integrated into the final ATRACO prototype located in the iSpace at the University of Essex. Afterwards we present the studies regarding dialogue strategies and discuss the outcome of the social evaluation. Finally, we present the results of the investigations regarding adaptive recognition, multitasking, and scalability that have been carried out as part of this work.

[1] Visit http://sourceforge.net/projects/owlspeak/ for further information.

Chapter 5
Experiments and Evaluation

In this chapter we report on the experiments and evaluation sessions we carried out with the OwlSpeak ASDM. During the implementation phase and the course of the initial investigations, we established an evaluation strategy that covers the three aspects: system integrity, dialogue optimisation, and practicability:

1. System integrity
   - Initial evaluation (Sect. 5.1)
   - System scalability (Sect. 5.2)
2. Dialogue optimisation
   - Topic switching strategies (Sect. 5.3)
   - Repair strategies (Sect. 5.4)
3. Practicability
   - Qualitative social evaluation (Sect. 5.5)
   - Advanced understanding methods (Sect. 5.6)

We focus on these topics since they cover the main functionalities that have been introduced in the previous chapters. The functional circle (Fig. 3.7) shows an overview on these functionalities. The multitasking capability, which is a requirement for *environment-centred adaptation*, has been tested as part of the initial evaluation that is presented in the following section. The initial system test also covers two main questions: is the ASDM stable and reliable and does the multitasking functionality work properly? Besides evaluation sessions with real users, we have investigated the technical capabilities of the OwlSpeak ASDM. A main matter is the scalability of the system since environments such as the ATRACO IE may consist of numerous devices and services that may be accessible via voice. Thus, in Sect. 5.2 the results of the scalability experiments in a large-scaled set-up are presented and analysed. This first part of our investigations motivated us to further analyse multitasking in SDS. In Sect. 5.3 we present the results of an in-depth study regarding the strategies that can be applied to facilitate a topic switch that

T. Heinroth and W. Minker, *Introducing Spoken Dialogue Systems into Intelligent Environments*, DOI 10.1007/978-1-4614-5383-3_5,

may occur during an ongoing human–system conversation. During this evaluation we compared the strategies presented in Sect. 3.6.2.1 in order to investigate their efficiency and user-friendliness. As defined in Fig. 3.7, topic switching refers to the aspect of *dialogue strategy adaptation*. To complete the investigation of this aspect, we present the results of an evaluation focussing on repair strategies in Sect. 5.4. To this effect we have analysed the repair strategies that are introduced in Sect. 3.6.2.2.

After the ASDM has proved its system integrity and its capabilities regarding dialogue optimisation, a qualitative social evaluation has been conducted within the framework of the ATRACO project. This study, presented in Sect. 5.5, aimed at the integratability of the system and at the multitasking features of OwlSpeak covering all aspects of environment-centred adaptation. We have also tested the capabilities of OwlSpeak regarding dialogue control, which is part of the *user-centred adaptation*. The second aspect of user-centred adaptation, speaker awareness, has been investigated in Schmitt et al. (2009). We have shown that OwlSpeak basically supports the main requirements to facilitate speaker aware dialogues. However, since emotional adaptation does not lie within the broader focus of this work, we refrain from evaluating speaker awareness but refer to Schmitt et al. (2011) for detailed information regarding this issue.

Motivated by the results of the qualitative study we have investigated methods to enhance the practicability of OwlSpeak. Thus, in Sect. 5.6 we present the results of an evaluation regarding the system's recognition and more specifically its understanding capabilities. Here, the main focus is set on the *intuitiveness* of the voice interface. This covers the first aspect of *SDS-centred adaptation*. The underlying idea was to find ways to improve the model-based approach to ASDM regarding the interpretation of user input. In the following section we present the initial system evaluation that has been conducted with the basic OwlSpeak ASDM (i.e. without any special dialogue strategies or the advanced understanding methods).

5.1 Initial Evaluation

We have implemented a set of complex spoken dialogues that were capable of influencing each other in order to test the stability and the reliability of the OwlSpeak ASDM. The underlying functionality has been defined in Sect. 4.3.1. The system utilises the scripting language that is described in Appendix C to exchange variables between different dialogue domains. This set-up allows to validate the main functionalities of the prototype. Furthermore, we were interested in the users' abilities to cope with one of the most important adaptation mechanisms: the switching between different dialogue tasks or topics. This functionality refers to spoken dialogue multitasking as described in Sect. 3.6.1. Here, a main challenge is the difficulty to *explicitly* evaluate the ASDM. Usually, users perceive the ASR and the TTS when interacting with an SDS. Hence, during an evaluation session, they rate the entire SDS when they are asked, for example, to rate the dialogue quality. For this reason we established a specific evaluation strategy that directly points at the multitasking capability of OwlSpeak. We divided the test field into two groups conducting the same spoken dialogues. However, each group utilises a different multitasking mechanism. In Sect. 4.2.3.3 we have introduced the static and the dynamic prioritisation of agendas to be carried out as dialogue acts. The first group of subjects conducted the dialogues by utilising the static adaptation strategy (Group A). The second group used the dynamic adaptation (Group B). Thus, both groups interact with the same SDS and practically have been engaged within the same tasks. These tasks, however, have been performed via different dialogue flows. This offers a way to estimate and therefore evaluate one of the most important system capabilities the ASDM offers: multitasking of different spoken dialogue tasks.

At the first attempt we tried to implement dialogues that are related to the application scenario and the "Prepare dinner" task as presented in Sect. 3.5. However, we experienced several issues with such a set-up due to the limitation of the test bed. To realise realistic dialogues for recipe selection and cooking support, several complex knowledge sources such as a database providing recipes, an "intelligent" fridge, several sensors in a kitchen, and user profiles of the people involved in the dinner would be required. To avoid this overhead we refrained from implementing the entire application scenario and decided to go for a scenario that has been investigated extensively in the area of SDM in the past: the travel booking domain (cf. Walker et al. 2002). On one hand, this scenario is sufficiently complex for testing the ASDM. On the other hand, it is not as demanding as the "Prepare dinner" task regarding its requirements. Notably, it is easier for the subjects to imagine that they must book a hotel and a flight compared to preparing a dinner without having access to an intelligent kitchen environment. Two main dialogues have been developed each as a separate SDO. The first dialogue provided the functionality to book a hotel. The second dialogue allowed flight booking according to the hotel stay or according to newly entered information. By utilising the concept of variables and the scripting language of OwlSpeak, it was possible to use the dialogues either combined for booking both flight and hotel or separately for accomplishing only one out of the two tasks. We have chosen this kind of dialogue layout so as to test the basic and the advanced dialogue management mechanisms of OwlSpeak.

5.1.1 Experimental Set-up

Besides the hotel and the flight SDO, we have defined a third SDO providing several reminders (cf. Sect. 5.1.1.1) for evaluating the multitasking capability of OwlSpeak. These reminders were used to simulate a topic switch and to investigate how users react on a spoken dialogue consisting of more than one task. As mentioned above, the subjects have been divided into two groups: Group A and Group B. The first group was asked to conclude the test run by serially processing each of the tasks to be accomplished: at the beginning book a hotel, then book an according flight, and finally listen to several reminders. In the following we refer to this approach as *static*. The latter group was confronted with a dialogue flow that dynamically mixed the tasks depending on the adaptive prioritisation algorithm presented in Sect. 4.2.3.3. We refer to this approach as *dynamic*. The main issue of the evaluation set-up was how to measure, which of the approaches performs better, and what does *better* mean in this context. The established SDS evaluation approaches such as SASSI (Hone and Graham 2000), PARADISE (Walker et al. 1997), or QUIS (Chin et al. 1988) are usually used to rate an entire SDS. Nevertheless, we refer to Wechsung and Naumann (2008) and decided to utilise SASSI for our evaluation sessions. It uses direct questions and was specifically developed for the evaluation of SDSs. Therefore it is more suitable than questionnaires developed for GUI-based systems such as the IBM Computer Usability Satisfaction Questionnaire (Lewis 1995). Objective metrics such as "task completion", "repetition rate", and "error rate" are also only partly useful for rating the multitasking capabilities of the ASDM.

Hence, we decided to evaluate this main adaptation-enabling aspect of the prototype by finding out which multitasking method is more convenient for the subjects. A suitable measure is the degree to that the subjects remembered the information after the dialogue terminated. Our aim is to find differences in, for example, the dialogue performance, the user behaviour, and the user-friendliness when the subjects use the dynamic or the static approach, respectively. Since the subjects were not explicitly informed about the fact that the aim is to keep the reminders in mind, we expected to receive a significant result about the usefulness of multitasking by dynamically mixing different spoken interaction tasks. Thus, the main question of the initial evaluation was: do users manage, i.e. are they able to perceive, more than one spoken dialogue task in parallel or is further assistance required if advanced SDS utilise comparable multitasking approaches within IEs? To complete the experimental set-up, we tailored the SASSI questionnaire so as to get an idea of the subjective user estimation regarding user-friendliness (see Sect. 5.1.1.2). In order to provide a realistic evaluation set-up we refrained from using a Wizard-of-Oz[1] layout. Instead we used an up-to-date English language SDS. The subjects utilised a SIP Voice-over-IP softphone running on a laptop providing microphone and speakers to connect to a SIP gateway.

This gateway is used to access a VoiceXML browser that interprets the dialogues generated by OwlSpeak. For the experimental set-up, we used the Voxeo Prophecy Server (Voxeo 2011), which provides an SIP gateway and a VoiceXML browser. The Nuance Recognizer 9.1 and the Nuance Vocalizer 5.1 have been connected via MRCP (Shanmugham et al. 2006). They perform the ASR and the TTS tasks. All dialogue descriptions (i.e. the SDOs) are in English language, thus all subjects were at least fluent or native speakers. In total 26 mixed-gender subjects aged between 25 and 50 years participated in the evaluation. Several users were experienced with SDS and with computers in general and some users only had little or no experience. However, since we asked for the user characteristics before the evaluation was

[1] A research experiment in which the subjects believe to interact with an autonomous computer system but in actuality interact with a faked system controlled by the unseen researchers.

Table 5.1 Exemplary dialogue excerpt that has been conducted during the initial system test

Speaker	Utterance	Dialogue	P_main	P_sub
	...			
System	How long do you want to stay?	D1		
Subject	What are my options?	D1	45	43,4
System	One week or two weeks.	D1		
Subject	One week.	D1	45	46,5
System	I should remind you to rent a movie for tonight.	D3		
System	You want to stay in the Hilton Garden Inn for one week. Your stay will begin on the 17th and end on the 24th. Is this correct?	D1		
Subject	Yes.	D1	45	–
System	Do you want to book a flight that corresponds to your time of stay?	D2		
Subject	Yes.	D2	44	35
	...			

started, we were able to equally distribute the subjects to the two groups. The first group was asked to process the dialogue using the static approach (Group A). The second group was asked to process the dialogue using the dynamic approach (Group B). Before the experiment started, the subjects received the following information:

- Try to imagine that you are talking to your IE, which is able to correspond with a travel agency so as to book a flight and a hotel for your next holiday.
- You plan to take a fictive leave starting from the 15th of August to the 29th of August.
- You want to travel to New York and stay there in the Hilton Garden Inn.
- If you do not know how to proceed with the dialogue you may always ask: "What are my options?"

After this introduction, the experiment started by conducting a system initiated dialogue that is explained in the following section.

5.1.1.1 Test Dialogues

We utilised three SDOs to encode the three dialogues used for the evaluation. Dialogue D1 described in the first ontology can be used to book a hotel, while dialogue D2 allows to book a flight that may correspond to the data the user has already provided during D1. The third ontology encodes dialogue D3. It consists of an initial greeting and three reminders that provide unrelated information. An extract of an exemplary dynamic dialogue flow is shown in Table 5.1. Two priorities are listed in Table 5.1: P_main and P_sub. The former value describes the priority of the two main dialogues D1 and D2.

Since D1 has a higher priority (45) it is started before D2 (with a priority of 44). Contrary to these static priorities, the reminders use a dynamic priority (P_sub). For each reminder this priority increases turn-wise. If P_sub is higher than P_main, the main dialogue is halted and the dialogue focusses on one of the reminders. Since the priorities of the reminders are lower than those of D1 and D2, Group A, which uses the static approach, receives them *after* D1 and D2 have been finished. The three reminders that are added to the main booking dialogues are:

- "Your friend Oliver has his birthday on the 15th of August. You may buy him a present."
- "I should remind you to rent a movie for tonight."
- "According to the weather report there will be heavy rain today."

In contrast, Group B receives the reminders dynamically during dialogue D1 and D2 depending on the individual duration of each dialogue step. The prioritisation methodologies are detailed in Sect. 4.2.3.3. In our example, the second reminder is stated during the hotel booking dialogue. Group A, who conducted the static dialogue, did not receive such interrupts but had to wait until the two main dialogues D1 and D2 were finished. In case the reminder would relate to information that gets more urgent (such as a pot of bowling water in the kitchen, etc.), the static approach would not be sufficient. Without using a different modality such as flashing lights to catch the user's attention, it is required to interrupt an ongoing dialogue in case of any urgent events. However, real-life conditions with realistic reminders could not be realised with our test bed. Thus, in our experiments, we relate to rather "artificial" statements used as reminders.

Since the flight and the hotel booking are encoded using two separate SDOs, it is also possible to book a flight that does not correspond to the data that have already been specified during the hotel booking. The system allows to implemented dialogues that can be used combined or separately. Besides the topic switching mechanisms, we also tested this technical benefit of the ASDM framework during the evaluation session.

5.1.1.2 Questionnaire

Upon completion of the dialogue interaction, the subjects were asked to fill in a questionnaire. It consists of four parts: personal data, familiarity with SDSs, comprehension questions, and questions that were related to the reminders. In the first part we ask for age, gender, English language skills, and for previous experience with SDS. The topic of the second part of the questionnaire focussed on the users' experience with the ASDM. The subjects had to rate the quality of the verbal communication, how the information has been conveyed, and how the dialogue flow has been perceived. Finally, the subjects were asked to rate how they liked or disliked the system.

As mentioned before, evaluating a Spoken Dialogue *Manager* is not a trivial task since a user always perceives and therefore rates the entire Spoken Dialogue *System* and not a specific and in this case crucial component of the SDS. Thus, the questions related to the overall experience can only be used to rate the SDS itself, which was not the main scope of the experiment. For measuring the performance of the ASDM, we refer to the third and the fourth part of the questionnaire: the comprehension and reminder specific questions. The following three questions ask for facts the subject were confronted with during the hotel and flight booking:

1. What is the name of the hotel you have booked?
2. At what time does your flight depart?
3. How much is the hotel room rate per night?

The last question is difficult to answer since the system mentions the price only once. The subjects were not really interested in the actual cost of the trip because of the artificial situation that could not be compared to a real holiday booking activity. Hence, they usually had problems answering this question correctly. The last five questions all relate to information provided by the reminders that were either added statically to the dialogue at the end or dynamically during the flight and the hotel booking:

1. Who has his birthday on which day?
2. What did the system remind you to do?
3. What is planned for the evening?
4. How will the weather be like?

The correct answers to these questions are:

1. Oliver on the 15th of August.
2. Rent a movie. Buy a present.
3. Watching a movie.
4. There will be rain.

A further aim of the initial system test was to reveal the type of multitasking spoken dialogues users prefer. Here, a main difference is the way the information is provided: task-wise in a static order or dynamically depending on the urgency of the upcoming information itself. In the following section we present the results of the initial system evaluation together with the outcome of the questionnaire analysis.

5.1.2 Evaluation Results

All participants concluded the dialogues successfully. They booked the correct hotel and a corresponding flight. Both groups were asked to rate the system on a scale from 1 (very bad) to 10 (very good) by answering a subset of the comprehensive SASSI questionnaire regarding the following quality factors:

EFF *Efficiency* indicates how efficient the system is, i.e. the test persons were asked to subjectively estimate how well they were able to follow the conversation.

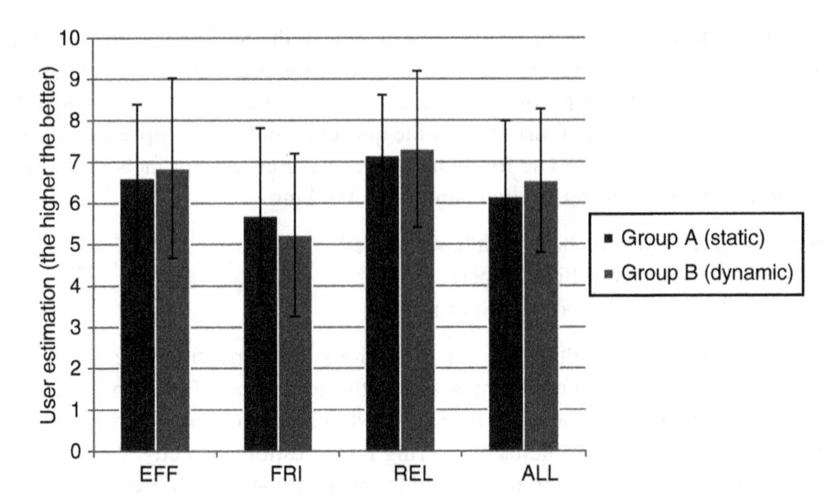

Fig. 5.1 The subjective usability estimations of Group A compared to Group B on average. The scales are efficiency (EFF), friendliness (FRI), reliability (REL), and an overall estimation (ALL)

FRI *Friendliness* describes how user-friendly the system is and how pleasant it is for the subject to interact with the system.

REL *Reliability* indicates how reliable the system is with respect to mistakes and understanding issues.

ALL *Overall* reflects an overall subjective estimation by the user of the entire SDS.

Figure 5.1 shows the questionnaire outcome of the subjective user estimation. The figure shows the mean values and the standard deviation of the ratings. In general, the SDS was rated positively by both groups independently of the applied approach. Overall, we see a tendency that the dynamic approach is slightly better rated by Group B regarding EFF, REL, and ALL. It seems that Group A considers the static approach to be more user-friendly. Notably, these values express tendencies. Due to the low number of test persons it was not possible to receive significant results supporting a hypothesis such as "the dynamic approach is better than the static one". In order to analyse the normal-distributed user scores we applied the analysis of variance (ANOVA—see Fahrmeir et al. 1984). We calculated P values to test for a significant difference between the two groups regarding the individual quality factors: EFF shows a P value of 0.779, FRI shows a P value of 0.49, REL shows a P value of 0.826, and ALL shows a P value of 0.603. Since all values are above the applied P threshold of 0.05, we cannot deduce an estimation regarding the benefits of the ASDM from these subjective measures. In order to investigate the tendencies, the subjective ratings showed, we carried out several objective measurements during the evaluation. In the following paragraph we present the objective results and investigate how they correlate with the subjective tendencies.

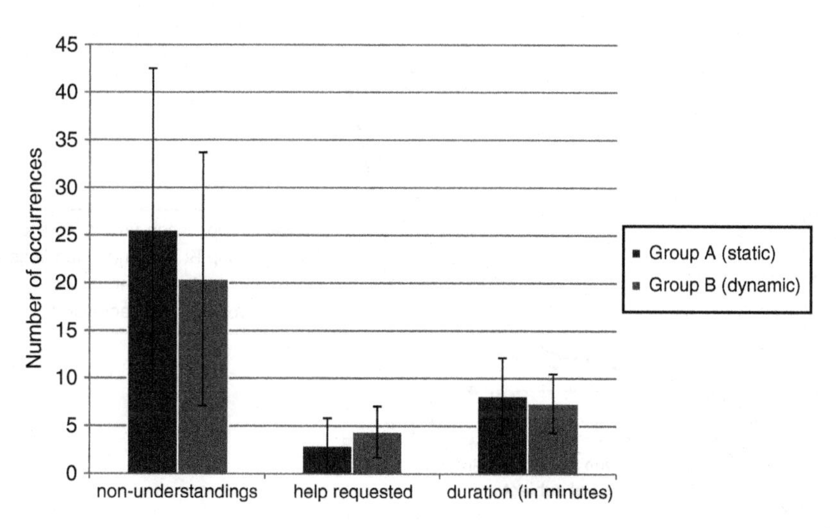

Fig. 5.2 The objective observations of Group A compared to Group B on average. The scales are the number of non-understandings, the number of times the users requested help, and the duration of the dialogues on average

Figure 5.2 shows how often the subjects requested help from the SDS, the number of non-understandings, and the required time duration to accomplish the dialogue. All scales show the average values and the positive and negative standard deviations. Notably, the two bars presenting the average number of non-understandings for the static and for the dynamic approach show a high standard deviation. For both groups we received wide ranging results from a minimum of three non-understandings to a maximum of 64 (Group A) and 45 (Group B), respectively. We assume these strong user-depended results to be caused by the combination of the English-language SDS and the non-native subjects. Nevertheless, as previously mentioned, all dialogues were successfully accomplished. Furthermore, we noticed three misunderstandings during all dialogues. The users had to repair the booking dialogue, i.e. book the flight or the hotel again, in case of such a misunderstanding. The values presented in Fig. 5.2 have also been analysed using ANOVA.

Again, due to the limited number of test persons we did not obtain significant results supporting an "is better than" assumption: the P value of the number of non-understandings is 0.416, the P value of the frequency, the users requested help is 0.211, and the P value of the dialogue durations is 0.602. As for the subjective measures only tendencies can be reported. The frequency of help requests is of interest in this context—it shows the lowest P value. We discuss this in the following. Two thirds of the non-understandings were caused by wrong or bad recognition: if the user had provided the specific entry via keyboard, the ASDM would have been able to react correctly. One third of the non-understandings were caused by input that has not been covered by the grammar. The user, for example, uttered "next Monday" while the system expected to receive "on the seventeenth".

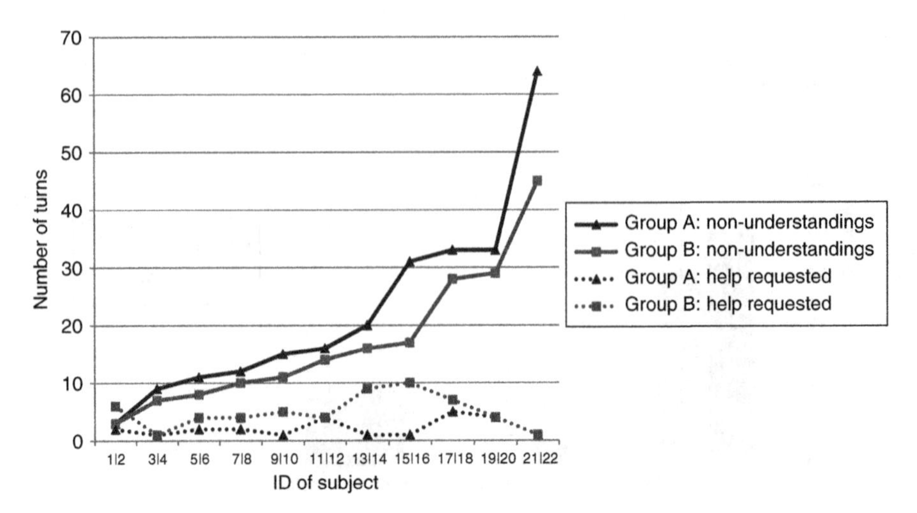

Fig. 5.3 The absolute number of non-understandings per subject and the corresponding number of requests for assistance

By requesting help from the system, the users were able to repair these situations. As a response, the ASDM generated a list of possible inputs the user may provide so as to make a correct request.

The objective results show some unexpected tendencies: on average, Group A received six more non-understanding situations compared to Group B. Furthermore, the subjects who used the static approach requested fewer help from the system—hence interacted less. Group A also needed slightly more time to accomplish the dialogue. Figure 5.3 underpins these results. The number of non-understandings and the frequency, a user requested help from the system are sorted by the number of non-understandings. We plotted all but four subjects in this way and investigated a correlation between the frequency of help requests and the number of non-understandings. Group A constantly receives more non-understandings while the subjects of Group B asked more or equally often for help. Furthermore, for the two third of the subjects who received less than 35 non-understandings (Group A and B combined), we constantly see less non-understandings in Group B while having more requests for assistance. These objective measures indicate that the Group B subjects, who used the more interactive dynamic approach, are more inclined to ask the system for assistance. Therefore they avoid high numbers of non-understandings and turn the entire dialogue more efficient.

However, our study also reveals a main drawback of the dynamic approach. The subjects were not explicitly told to take care of any additional information but only to complete the two "travel booking" tasks. Thus, we expected that only a few test persons would remember all information provided by the system. Since Group A received the three reminders *after* the main tasks have been concluded, we expected that Group A would experience a slight advantage. To allow rating

Table 5.2 The number of subjects per group and per category who answered correctly

	Group A (static)	Group B (dynamic)	Group A+B
Friend's name	0 (0%)	0 (0%)	0 (0%)
Day of Birth	8 (62%)	5 (38%)	13 (50%)
Rent movie	6 (46%)	2 (15%)	8 (31%)
Buying present	9 (69%)	3 (23%)	12 (46%)
Watching movie	6 (46%)	2 (15%)	8 (31%)
Rainy weather	8 (62%)	9 (69%)	17 (65%)

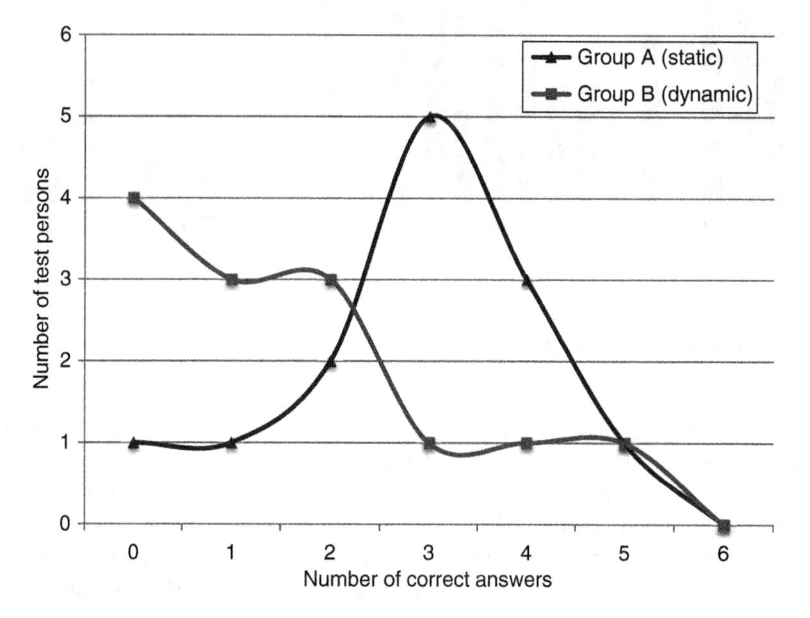

Fig. 5.4 The number of subjects related to the number of correct answers

the outcome of the questionnaire, we counted one point for each topic the specific subject remembered correctly. Thus, a subject who did not remember anything that has been provided as part of the reminders would gain zero points. In contrast, a test person who memorised all the provided information would gain six points. One for the correct name of the friend, one for the right date of birth, one for renting a movie, one for buying a present, one for watching a movie, and a further one for remembering the rainy weather. Table 5.2 shows the number of subjects per group who answered correctly. We assume that the name of the friend—Oliver—was only poorly synthesised, thus no one was able to keep this information in mind.

The static group outperforms the dynamic one in all but one category. This last category, the information about the weather, was provided at the end of both the static and the dynamic dialogues. The subjects of Group B perceived less information. If we take a closer look at how the numbers of correct answers are distributed amongst the subjects, the discrepancy between Group A and Group B can be revealed. Figure 5.4 shows the distribution of correct answers.

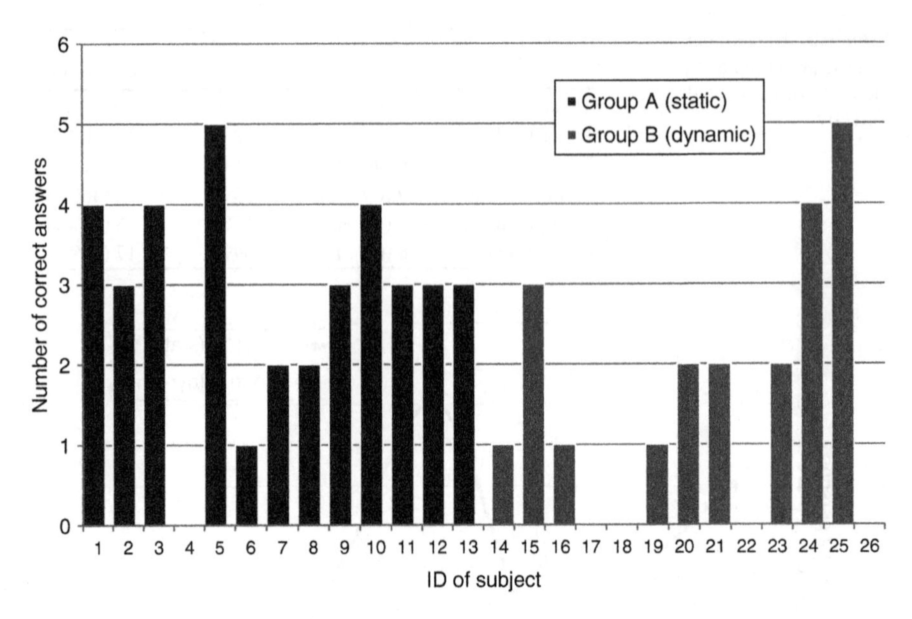

Fig. 5.5 The number of correct answers per user

Notably, the frequency of correct answers in Group A is nearly Gaussian distributed around the average value of three right answers. Contrary to this, the highest number of Group B subjects remembered not a single item of information the system provided. Expressed in absolute numbers, only three out of thirteen Group B subjects answered more than two questions correctly. Figure 5.5 shows the number of correct answers for all subjects. The Mann–Whitney U-Test calculates a P value of 0.039 underpinning the significance of the differences between the two approaches (Mann and Whitney 1947). This indicates that the members of Group B did not perceive the provided information in the same way the other subjects did. A reason for this may be the fact that the test persons were nearly totally occupied by managing the hotel and flight booking task. Since the system did not provide any assistance in switching to a "reminder" task, the users blanked out the information that was not related to the main goal: the "booking" task.

The results that have been analysed indicate a tendency towards a more "user-involving" dynamic dialogue on the one hand and a more efficient static dialogue on the other hand. During a free discussion that followed each experiment, several participants stated that they had consciously ignored all information that did not directly relate to the main tasks. However, if the subjects had been informed beforehand that they will have to answer specific questions about the dialogue history, we probably would have obtained a totally different result. Certainly, it would be far from practice if users were not confronted with such reminders spontaneously. Overall the evaluation revealed insights into the users' behaviour when confronted with an SDS that behaves adaptively.

5.1.3 Conclusion

The version of the OwlSpeak ASDM that has been tested during the initial evaluation series fulfils several requirements that arise from the multiple tasks-based situations that occur within IEs. The ASDM is able to pause and resume active tasks, add and remove dialogue domains, permanently save the state of a dialogue, and furthermore it can provide more than one active spoken dialogue in parallel. These functionalities allow for adaptive spoken dialogues that cover several disjoint or partly overlapping domains. However, from the user's point of view it is a different assignment to solve a single task (e.g. book a flight) or to interface with an IE that provides a variety of controls and tasks via speech. Hence, the main question that must be answered before multitasking SDSs will be extensively used in everyday life is *how do users cope with the multitasking capabilities of an SDS* such as the proposed prototype provides? The evaluation revealed some interesting insights: on one hand, the dynamic approach encourages the users to be more responsive to the SDS (i.e. ask for assistance) and thus they completed the dialogue with fewer non-understandings. On the other hand, most Group B subjects memorised less facts than the Group A subjects, making the dialogue less effective.

Compared to GUIs, the users are accustomed to multitasking capabilities: techniques such as taskbars or widgets are adopted by the users. However, when it comes to spoken interaction we are far away from such a widespread user acceptance. Enhancing the common usage of SDSs by adding the functionality of multitasking could be an important step towards wider application of spoken interfaces. The evaluation revealed several questions that must be answered beforehand. One of the most important questions related to multitasking is *how can a focus switch from one task to another be signalised by the system and/or by the user?* We investigate this question in Sect. 5.3 and present the results of a further user evaluation that has been motivated by the results of the initial experiments. In the following section we discuss the scalability of the OwlSpeak ASDM, in order to demonstrate the system integrity.

5.2 Scalability Experiments

In this section we present the results of the scalability analysis that has been conducted in order to proof the practical application of the OwlSpeak ASDM. During the initial system evaluation we tested comprehensive SDOs for flight booking (number of OWL individuals > 100) and for hotel reservation (number of OWL individuals > 200). The results showed that the framework is able to handle complex dialogues. For the set-up of the social evaluation in the iSpace it was necessary to develop a set of lightweight SDOs for specific devices with a limited number of functionalities that should be controlled via speech. During this evaluation series the ASDM controlled a maximum of six devices and services due to the limitations of the environment. Thus, we only proved the scalability of the system regarding the number of OWL individuals (from simple to complex dialogues) but not regarding the number of dialogue models (SDOs), the ASDM must incorporate into the dialogue.

5.2.1 Experimental Set-up

This scalability analysis investigates how the ASDM performs regarding metrics such as the average start-up duration and the latency within a command-and-control scenario where the number of devices scales up to 100. For the experiment we developed 100 lighting control ontologies with a similar set of commands as outlined in Appendix A.3. Each SDO consists of 30 OWL individuals for the grammars, the utterances, the semantic values, and system related objects such as the user workspace. We started with one SDO and protocolled the duration of the system start, the latency of a dialogue request, and the latency of a "work" command. The dialogue request is the main function of the ASDM that is called when a dialogue description, for example in VoiceXML, is requested by the SDS. The "work" command is called whenever the user provides a specific input (see Table 4.1 in Sect. 4.2.1). Both commands are vital to the dialogue generation and control. We subsequently added a further SDO and ran the analysis again. We completed the analysis after the maximum number of 100 devices has been reached. We assume that it is not required to control more devices and services via voice within a specific IE in practice.

For measuring the scalability of the ASDM we took the system reaction time into account. Notably, the time the user needs to enter a spoken command is not considered. The hardware and the software set-up of the scalability analysis were comparable to the set-up that has been used during the other evaluation series. The following software components and frameworks have been used by the OwlSpeak ASDM:

- JDOM 1.1.2 is used as a Java-based solution for accessing, manipulating, and outputting XML data from Java code.
- The Java Servlet API 2.5 is used as a standard framework to provide Servlets that generate the dialogue output.
- The OWL API 3.2.4 is used to interface with the SDOs (i.e. to read and manipulate the knowledgebase). This API is the basis of our framework that facilitate the access and the creation of SDO-specific OWL individuals.
- Jetty 7.5.4 is used as Servlet container and web server. It allows the SDS to communicate with the ASDM via HTTP.
- The ASDM runs on the 64 Bit Windows 7 operating system.

The ASDM itself was installed together with the complete set of all SDOs on a workstation machine (Intel Core2Quad Q6700 with 8GB of RAM). In the following we provide the results of the scalability analysis before we discuss the outcome of the experiment.

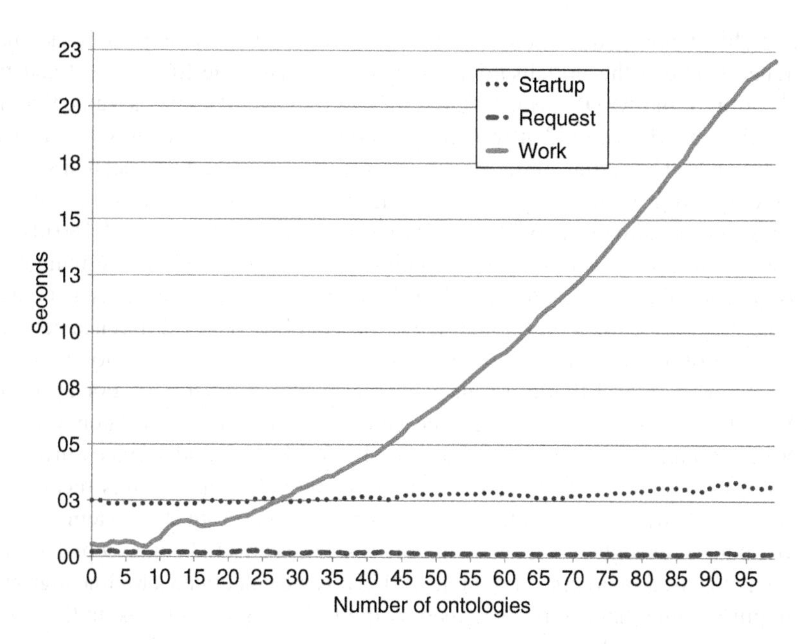

Fig. 5.6 The duration of the initial system startup, of a request, and of a work command for a varying number of Spoken Dialogue Ontologies involved in the dialogue generation

5.2.2 *Evaluation Results*

We decided to investigate the performance of the ASDM by focussing on three metrics that are described in the following:

Startup The *Startup* metric is the time duration in seconds, the ASDM requires to initialise all SDOs (via the OWL API) and to start the Servlets running on the Jetty Servlet Container. Thus, the boot up duration of the ASDM includes the duration Jetty needs for initialisation.

Request The *Request* metric is the time duration in seconds, the ASDM requires to generate a new output that provides the spoken dialogue. No input by the user and no output by the system must be processed for this functionality. If new SDOs (i.e. new devices, services, or dialogues) are incorporated during runtime, the ASDM would provide a modified dialogue.

Work The *Work* metric is the time duration in seconds, the ASDM requires to process an input by the user or an output on part of the system. In case of a user input, the system must check if there are any ambiguities or if the user command can be executed immediately.

Figure 5.6 plots the outcome of the scalability analysis. The *Startup* duration of the ASDM proved to be robust against the number of SDOs involved in dialogue generation. On average the system needed 2.7 s for the initialisation.

A slight standard deviation of 264 ms indicates that there is only a small difference between the initialisation of the system with one SDO (2.5 s) and the initialisation of the ASDM with 100 SDOs (3.2 s). In summary, we assume that the boot up time duration would grow similarly slowly if the experiment was scaled up to 1,000 SDOs. The results of the *Request* metric also show that the mechanisms we implemented to provide newly generated dialogues are robust against the number of SDOs that are involved. We have measured an average duration of 180 ms per request. Again a very low standard deviation (35 ms) indicates that the number of SDOs do not influence the time, the ASDM requires to provide a newly generated dialogue. Regarding the *Request* metric, we did not observe any difference between the duration of a request command with one SDO and the time duration with 100 SDOs. A completely different result is shown by the *Work* metric. For up to ten SDOs that are involved in the dialogue generation, we observe a latency of the ASDM less than 1 s. However, if there are 10–30 devices and services involved, the ASDM needs up to 3 s to process the user input. In practice, this means that the user has to wait for 3 s until he receives a response from the system or until his command is executed. A latency of more than 3 s seems to be unacceptable for a system that is used in practice. For up to 100 SDOs we calculated a loglinear growth of computing time that can be described by the following function using the big $\mathcal{O}$ notation (Bachmann 1894):

$$\mathcal{O}\left(\left(\frac{n}{k}\right) \cdot \log_2(n)\right)$$

with n is the number of SDOs and a constant factor $k = 2$. We calculated this factor to describe the reduction of the results that we measured. A main reason for the increasing computing time is the ambiguity detection that is part of the ASDM. This detection is carried out each time the user provides input. Ambiguities are being detected by *pairwise* comparing each grammar string of each grammar individual of all SDOs that are currently active. The current state of the prototype does not allow sub-grammars. This restriction allows for a simplified comparison of the grammars. Without this constraint we assume that for huge numbers of SDOs ($n > 100,000$) the computational time will be within $\mathcal{O}(n^2)$. Obviously, this is a only a theoretical limitation of the ASDM—in practice such huge numbers of devices and services would not be reached. Nevertheless, the comparison of complex grammars (and most notably of sub-grammars) is an issue within a scaled-up version of the system. In the following section we conclude the scalability report before we focus on dialogue optimisation in Sect. 5.3.

5.2.3 Conclusion

In this section we presented the scalability analysis results obtained from the OwlSpeak ASDM. The experiment revealed that both the required time duration for initialisation and the time the ASDM needs to provide a newly generated dialogue are robust against large numbers of SDOs. However, the processing of spoken user input is computational more demanding since an ambiguity check is carried out for each user turn that is conducted. The pair-wise comparison leads to a worst case complexity of $\mathcal{O}(n^2)$, which is too high for numerous SDOs (i.e. devices and services). A fluent dialogue would not be possible anymore. In practice, we do not assume this to be an issue. Usually, a voice interface, as the ASDM provides, would grant access to 10–20 devices and services. These numbers do not lead to high latencies. If there are, for any reasons, higher numbers of SDOs involved in the dialogue generation, a preceding SDO selection would be required. Such an SDO selection, for example, based on semantic domain detection (see Sect. 4.2.3.4) would be applied before the grammars are checked for ambiguities in order to keep the number of possible dialogue domains low.

5.3 Topic Switching Strategies

After the investigations regarding the system integrity, we present in this section the results of an evaluation focussing on dialogue optimisation. We investigate topic switching strategies that assist the user. Our hypothesis is that such strategies compensate the drawbacks caused by the dynamic dialogue switching we have faced during the initial system test. In Sect. 3.6.2.1 we presented three approaches that support dialogue switching by providing discourse markers, task explanations, and the option to negotiate. In the following we compared these strategies (*discourse*, *explanation*, and *negotiation*) with the dynamic dialogue switch (referred to as *baseline*). The baseline approach has been tested as part of the initial system evaluation; it runs without any assistance.

5.3.1 Experimental Set-up

As for the initial experiments we refrained from using a Wizard-of-Oz layout in order to guarantee a realistic evaluation set-up. For this evaluation, however, we used an up-to-date German language SDS. Due to the modular system architecture of OwlSpeak it is possible to replace the English language system by a German SDS. We were able to recruit more subjects since English skills are not required. Our main goal was to investigate the drawbacks and the benefits of the presented task switching strategies. For this reason we deployed a system with a performance (regarding recognition rate etc.) that is comparable to established systems. Figure 5.7 shows an architectural overview on the SDS that utilises OwlSpeak as core ASDM component. The subjects used a SIP phone running on a laptop providing microphone and speakers to connect to a SIP gateway that grants access to the SDS. For the experimental set-up we used Voxeo Prophecy for providing the required SIP gateway and an appropriate VoiceXML browser. The Nuance Recognizer 9.1 and Nuance Vocalizer 5.1 are accessible via MRCP (Shanmugham et al. 2006). They provide the ASR and the TTS functionalities.

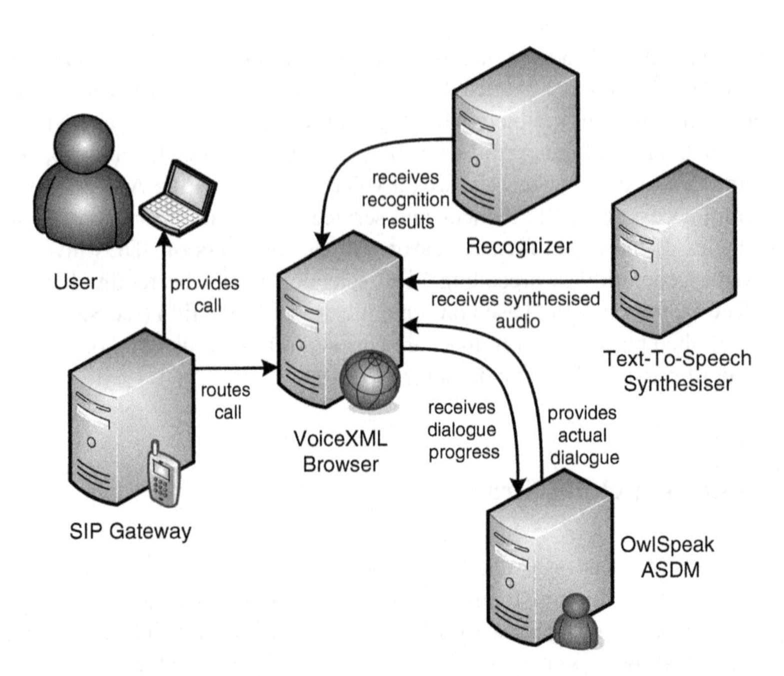

Fig. 5.7 Architectural set-up of the OwlSpeak-based SDS used for the evaluation

All dialogue descriptions (i.e. the SDOs as described in Sect. 4.3) are in German language since all subjects were fluent or native speakers. In total 80 persons took part of the experiment. A majority of the subjects aged between 19 and 47 (25.5 on average) were not experienced with SDSs but were familiar with computers in general. The 26 women and the 54 men were separated into 4 groups with 20 people each so as to equally cover the three interruption strategies and the baseline set-up. Figure 5.8 shows how the subjects are distributed to the four groups. While Groups A, B, and C are nearly equal in age and gender, the negotiation group consists of slightly more male than female users and, on average, is 3 years older. We do not estimate that this influences the results of the experiment. Analogously to the initial evaluation, the subjects' default task was to make a reservation for a hotel in New York and an according flight to the USA. This booking task was interrupted by a short reminder, a long reminder, and a sub-dialogue. No information about the interrupting tasks was given in advance to the user in order to examine his reaction and situational awareness similarly to real-life processes. We provided the subjects with the same information as for the initial evaluation:

- Imagine that you are talking to your IE, which is able to connect to a travel agency in order to book a flight and a hotel for your next holiday.
- You plan to take a fictive leave starting from the 15th of August to the 29th of August.
- You want to travel to New York and stay there in the Hilton Garden Inn.
- If you do not know how to proceed with the dialogue you may always ask: "What are my options?"

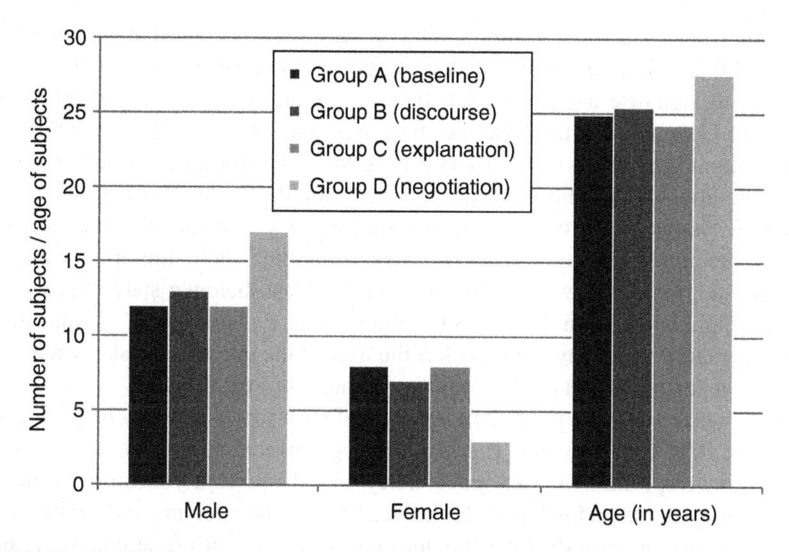

Fig. 5.8 The distribution of male and female subjects and their average age

A benefit of keeping the set-up of the experiments identical is that the result can be compared. However, the re-design of the SDS to a German language system may influence the results. This was the main reason why we decided to allocate 20 subjects to run a baseline test. In the following section we provide examples of the test dialogues and of the interrupting tasks. Furthermore, we explain the benefits of the three tasks switching strategies, we estimated.

5.3.1.1 Test Dialogues

We introduced three strategies that may facilitate task switches in Sect. 3.6.2.1. These were integrated into a travel booking task as presented in Sect. 5.1. While the test dialogue itself was similar to the example presented in Table 5.1, we modified the content of the interrupting tasks so as to apply the topic switching strategies. The three interruptions are:

Short reminder "The washing machine is done." This short reminder was provided during the first part of the test dialogue.

Long reminder "Your friend Mario celebrates his birthday on Thursday. You may buy him a present." This long reminder was provided during the middle part of the dialogue.

Sub-dialogue "You are supposed to meet Sandra for dinner on the 10th of March. Do you want to make a reservation? [...] You can choose between *San Marco*, *Enchilada* and *Chinese World*." This complex sub-dialogue was provided during the last part of the dialogue.

These interrupting tasks were combined with the three strategies that we introduced in Sect. 3.6.2.1. The *discourse* strategy applies discourse markers such as “ah” and “so” to signalise a focus switch from the main task to the interrupting one and back. Compared to the baseline, the discourse strategy intended for raising the situational awareness of the users during the multitasking dialogue. However, the ambiguities of specific discourse markers could also lead to confusion. The *explanation* strategy introduces the interrupting task with an utterance such as “excuse me for interrupting you” or “now something more important” and re-initialises the main task with a description of the last dialogue state: “let’s go on with the flight reservation. You selected Stuttgart as a start airport.” Similar to the discourse markers the long phrases alert the user of the upcoming task switch. This may have a positive effect on the cognitive resources of the subjects.

On the other hand, the precise explanation of each topic shift may be boring and even frustrating for the users. The third strategy, *negotiation*, asks the user if he wants to interrupt the main dialogue: “sorry, may I interrupt you for a moment?” The user may agree and listen to the interruption or may decline and suppress the subtask. In case the user declines, he does not receive the information. We assume that within a real-life situation the interruptions are time-critical. For this reason, it would not be realistic to receive the information after the dialogue has terminated. The ability of the user to directly controlling the dialogue may result in higher user acceptance and situation awareness. As mentioned, an essential disadvantage of the negotiation strategy is that some critically important tasks may be rejected. Similar to the initial system test, Group A received the interruptions without any applied strategy. Group B received the discourse markers, Group C the explanations, and Group D the negotiative strategy. In the following section we discuss the questionnaire before we focus on the results of the experiment.

5.3.1.2 Questionnaire

After having successfully accomplished the dialogue the test persons were asked to fill in a questionnaire related to the main task of the dialogue and to the interrupting tasks. Again, we adopted the SASSI method, as we did for the initial evaluation (Hone and Graham 2000). We have tailored the comprehensive questionnaire to 10 questions related to the following quality factors:

EFF *Efficiency* shows the efficiency of the system, i.e. how well the dialogue flow can be comprehended by the user.

FRI *Friendliness* describes how user-friendly the system is and how pleasant it is for the subjects to interact with the system.

REL *Reliability* shows the reliability of the system regarding mistakes and understanding problems.

NOTIRR *Not irritating* presents the degree to that the user estimates that he is not irritated or confused by the task switches.

CLS *Clearly separated* describes the subject's ability to separate the interrupting tasks from the main dialogue.

The subjects used checkboxes with values corresponding to an evaluation scale from one (very bad / strong disagreement) to five (very good / strong agreement) to rate the system. In order to derive a valid estimation about the topic switching strategies, the subjects were asked to answer several comprehension questions regarding the content of the three reminders and the content of the main booking dialogue. We collected this information in order to be able to estimate how the users perceived the information of both main task *and* of the interruptions. The questions regarding the main task are:

1. What is the name of the hotel you have booked?
2. How long will you stay abroad?
3. From which airport will you depart in Germany?

The correct answers to these questions depend on the input of the user during the experiment. We have recorded all dialogues to allow doublechecking the correctness of the answers. The following four questions relate to the interrupting tasks:

1. Of what kind of domestic appliance have you been reminded?
2. Who celebrates his birthday?
3. When is the birthday you have been reminded of?
4. With whom do you have an appointment for dinner?

The correct answers to these questions are:

1. A washing machine.
2. Mario.
3. On Thursday.
4. With Sandra.

Similarly to the initial system test, we awarded points for correct answers when analysing the collected data. The results and their interpretation are detailed in the following section.

5.3.2 Evaluation Results

Figure 5.9 plots the subjective results of the questionnaire on average. During all dialogues, the main tasks of booking a hotel and a flight have been accomplished successfully. For each quality factor a comparison of the four task switching strategies is presented. Regarding FRI and EFF no significant differences can be recognised. We applied the Kruskal–Wallis test (Kruskal and Wallis 1952) for testing if a quality factor of a specific group is (statistically) significantly different compared to the same quality factor of another group. For FRI we calculated a P value of 0.361 and for EFF a P value of 0.146, respectively. These values indicate that we cannot deduce anything from the comparison of these quality factors.

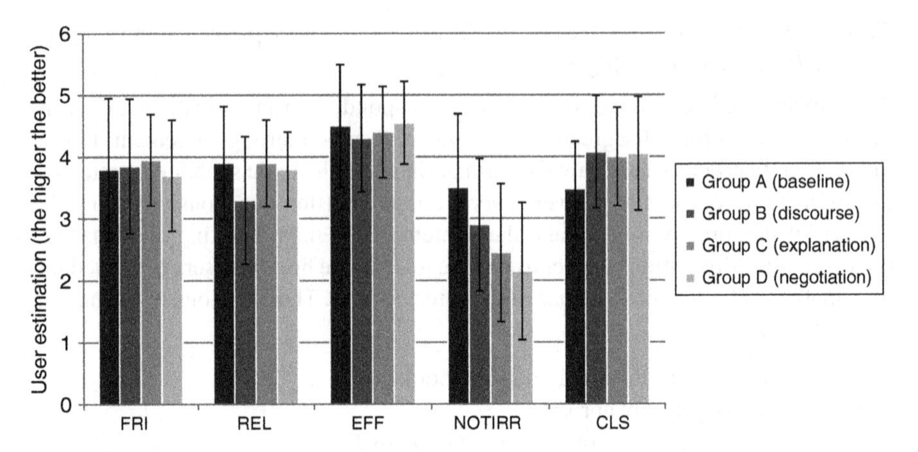

Fig. 5.9 The five combined quality factor scores

On average and independently from the applied strategy, all subjects rated the system between "neutral" and "good" regarding user friendliness and between "good" and "very good" regarding efficiency. The same test deployed on REL, NOTIRR, and CLS revealed significant differences. Regarding the REL factor of Group B compared to the other groups we calculated a P value of 0.019. The NOTIRR factor significantly differs for all groups (P value < 0.005). The CLS factor shows a significant difference when Group A is compared with the other groups (P value $= 0.013$). The low P value of the REL factor indicates that the subjects of Group B rated the *discourse* strategy significantly lower regarding the reliability of the system compared to the subjects who used the other strategies. For all groups, the Kruskal–Wallis test shows a significant difference between the NOTIRR factors. Group A estimates the baseline approach to be the least irritating one, while the subjects who used the *negotiation* strategy (Group D) rated their approach to be the most irritating one.

A further significant result is the difference between the *baseline* strategy and the other approaches. The subjects of Group A agree that the topic switch without any assistance by the system does not clearly separate the interrupting tasks from the main dialogue. The other groups had fewer issues with differentiating between the tasks. This finding underpins the motivation of the evaluation presented in this section. In summary, the questionnaire revealed *baseline* to be the least irritating strategy for the users. However, it also reveals that *baseline* is not suitable for separating the interrupting tasks from the main dialogue. This separation is an important requirement for performing efficient topic switching. We presuppose that, regarding the interruption, users would remember the provided information more easily if they perceive that *the interruption happened at all.* The *discourse* strategy was perceived to be the least reliable one.

A reason could be that the colloquial discourse markers and the less formal mode of speaking lead to the estimation that the system is not reliable. The *explanation*

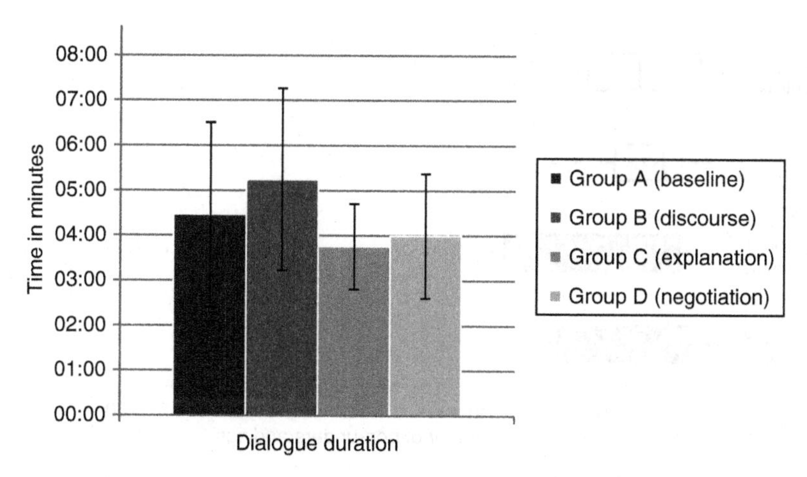

Fig. 5.10 The average duration of the dialogues in minutes and the standard deviation

and the *negotiation* strategies are equally rated. Only regarding the NOTIRR factor *explanation* is significantly better rated than *negotiation*. We were surprised by the bad rating of the more sophisticated approaches. After the first test runs, when this result started to stand out, we decided to ask the subjects for the reason why the approaches were rated to be irritating. All answers pointed into the same direction: the *baseline* approach is not irritating since the user's aim was to arrange the booking tasks—the users *did not want* to listen to any interruptions at all. Thus, the approaches that assisted the subjects to better perceive the interruptions were rated as irritating. Besides the subjective results we also collected several objective values so as to further investigate the impact of topic switching strategies on the dialogue and on the user. Figure 5.10 plots the average dialogue duration for the four groups.

While Group A, using the *baseline* approach, and Group B, using the *discourse* strategy, needed more than 4 min for accomplishing the dialogue on average, Group C and Group D performed significantly faster. The average values visualise a tendency that the *discourse* strategy performs worse than the other approaches. This tendency turns out to be significant: the analysis of variance (ANOVA—see Fahrmeir et al. 1984) calculates a P value of 0.05 underpinning this hypothesis. Furthermore, a pairwise difference between Groups A + B and Groups C + D also reveals to be significant: the Student's t-test (Mankiewicz 2000) shows a P value of 0.019 indicating that on average, compared to the *baseline* and the *discourse* strategy, Group C and Group D accomplished the dialogue significantly faster. However, the quickest and the lengthiest dialogues were protocolled as part of the Group B session. One of the subjects only needed 2:22 min to complete the dialogue while a subject of the same group needed 10:52 min. Besides the duration of the dialogue, we also measured the number of non-understandings that occurred during a dialogue.

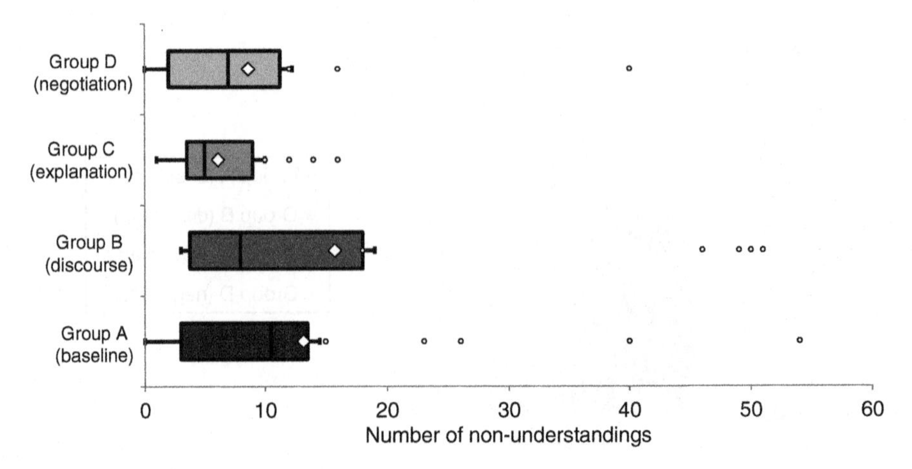

Fig. 5.11 The distribution of non-understandings for the four groups plotted as a Box–Whisker chart

Since we have received widespread results regarding this factor, we have plotted them in a Box–Whisker chart. Figure 5.11 shows the distribution of the non-understandings for the four groups. Starting with Group A, the plot indicates that 75% of the subjects received less than 13 non-understandings. This result is slightly better than the results of the initial evaluation. We assume that such an improvement of performance is due to the usage of a German SDS. However, for Group A, the chart also shows five outliners who received a number of non-understandings ranging from 15 to 54. Group B, on average, shows the highest number of non-understandings. Seventy-five per cent of the Group B subjects received between 3 and 19 non-understandings. We have also recorded four outliners between 45 and 52 non-understandings. Group C shows the best result regarding the average and the median. Seventy-five per cent of the subjects who used the *explanation* strategy were confronted with less than eight non-understandings. The outliners of this group are within the range of the third quartile of Group B: between 11 and 16. Group D has a higher median and average value than Group C but still outperforms Group A and B. Seventy-five per cent of the Group D subjects received less than 12 non-understandings. In this group we have three outliners with 16 and one outliner who received 40 non-understandings.

The plot indicates that the *explanation* strategy outperforms the other strategies in terms of the number of non-understandings. However, this assumption does not seem to be statistically significant: the analysis of variance calculates a P value of 0.066. Thus, we tested for a pairwise difference between Groups A + B and Group C + D. Here the Student's t-test reveals a significant difference ($P < 0.012$). Both objective measures indicate that C and D perform significantly better than A and B regarding the dialogue duration and the number of non-understandings. These values both relate to the efficiency of a conversation, which undoubtedly is

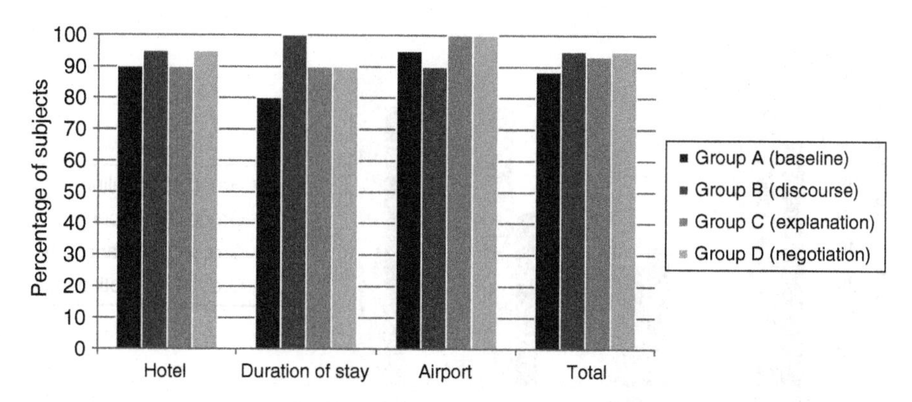

Fig. 5.12 The average percentage of what the subjects kept in mind regarding the main dialogue for the four groups

an important system property. In the following we further investigate the objective metrics by analysing the information, the subjects remembered after the dialogue was completed.

Figure 5.12 depicts the average percentage of information regarding the main dialogue, the user was able to remember. All subjects remembered at least 80% of the provided information we asked for. For all groups 90–95% of the subjects remembered the name of the hotel they have booked correctly. For several subjects it was difficult to recall the duration of the stay: on average only 90% of all subjects remembered the duration correctly. Nevertheless, most of the subjects (96%) were able to correctly indicate the departure airport. We noticed a slightly worse result of Group A when comparing all answers of the subjects on average (see "Total" bars in Fig. 5.12). However, this result does not seem to be statistically significant—we can infer that all subjects of all groups were concentrated and keenly followed the dialogues. This user attention is important since we intend to use the degree of what the subjects correctly remembered regarding the interrupting tasks as an efficiency metric of the topic switching strategies. A logic requirement for such an investigation is that all groups equally follow and equally perceive the dialogue. In the following we analyse the results of the questions related to the interrupting tasks.

Figure 5.13 depicts the average percentage of subjects who were aware of a specific interrupting task. The difference between the four groups are significant: the analysis of variance calculates a P value of 0.002 indicating that the different strategies had a strong influence on the way the users perceived the information. The short reminder shows very poor results both for the *baseline* and the *discourse* strategies. In Group C only 70% of the subjects noticed this interruption. The *negotiation* approach shows a result above 90%. However, only 18 out of the 20 subjects allowed this interruption. Again, a P value of 0.002 indicates that the differences between the groups are significant regarding the short reminder. Group A performed similarly regarding the long reminder. The Group B subjects performed significantly better compared to the short reminder.

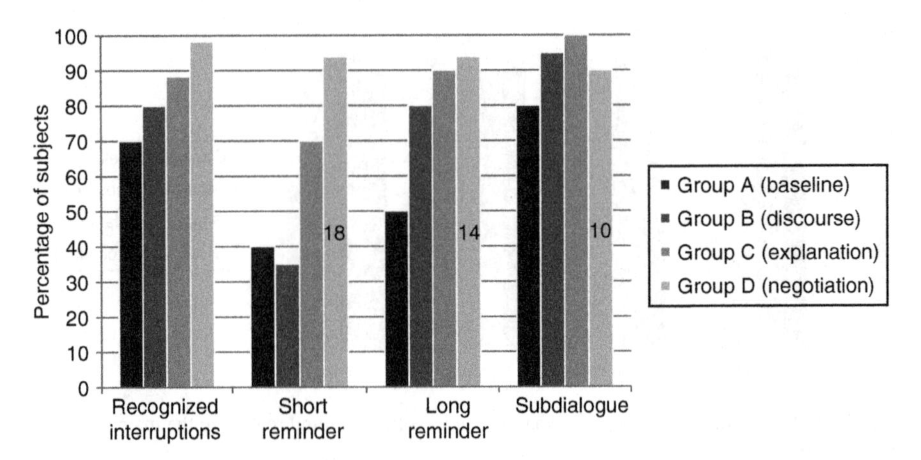

Fig. 5.13 The average percentage of interruptions the subjects have been aware of. The first group of bars indicates that the subjects have noticed there was an interruption at all. The other bars indicate of what interruption-topics they have been aware of. The numbers on the bars of Group D indicate how many subjects allowed the respective interruption

Furthermore, 90% of the subjects in Group C recalled the long reminder. On average, this result is only surpassed by the *negotiation* strategy. However, only 14 out of the 20 subjects admitted this interruption. These results are also statistically significant: we calculate a P value of 0.028 underpinning the observation of the average values. The last group of bars depicted in Fig. 5.13 shows the percentage of users who noticed the sub-dialogue. Since this was the last subtask that occurred during the dialogue most of the users were aware of this interruption. Again, the *baseline* approach performed worst: 80% of the subjects noticed this interruption. All subjects in Group C recognised the sub-dialogue. This result outperforms the *negotiation* strategy: 90% of the ten subjects who allowed this interruption remembered that there was a sub-dialogue. As for the previous results we calculated a P value below 0.05 indicating that the differences are statistically significant.

The main goal of the comprehension questions that were part of the questionnaire was to determine the ability of the subjects to memorise the details of the reminders. We have expected that, depending on the applied strategy, the users do not remember similar pieces of information. The results are shown in Fig. 5.14. Again, the four interrupting strategies have been compared. The y-axis shows the percentage of correct user answers to the detailed questions. Regarding the short reminder we asked for the name of the home appliance (washing machine). Regarding the long reminder we asked for the name of the friend who celebrates his birthday (Mario) and the date of his birthday (on Thursday). Regarding the sub-dialogue we asked for the name of the person who dates you for dinner (Sandra). Since the results we received were not normally distributed, we applied the Kruskal–Wallis test to check for significant results. The answers to the first question led to a two-part test field: Group C and D perform significantly better than Group A and B (P value $= 0.007$).

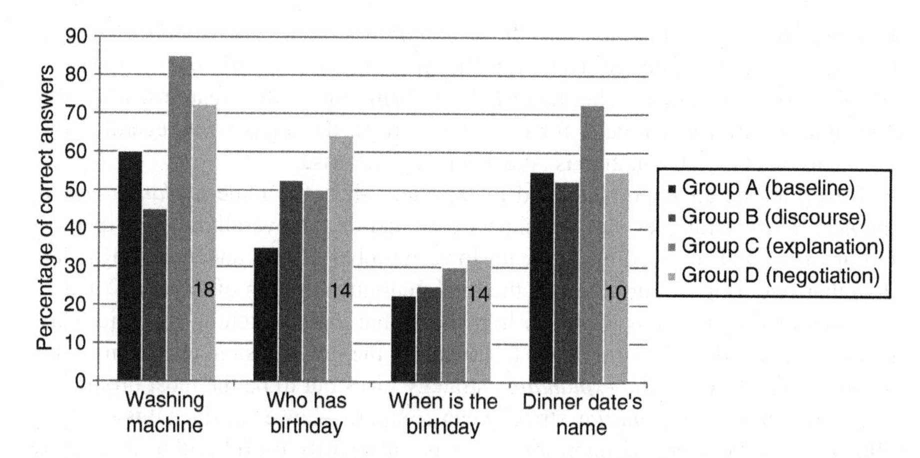

Fig. 5.14 The average percentage of what the subjects kept in mind regarding the interrupting tasks. The numbers on the bars of Group D indicate how many subjects allowed the respective interruption

This result correlates with the analysis of the objective data. During this part of the evaluation, Group B performed worst. The birthday reminder turned out to be the most difficult information.

The differences between the values we have collected regarding the long interruption are not significant. Thus, we can estimate a tendency that the *explanation* strategy does not perform worse than the other strategies. However, the answers to the last questions reveal a significant result. Over 70% of the answers of Group C were correct. The other groups are only slightly above 50%. Group C, who applied the *explanation* strategy, showed the highest values of correctly remembered details regarding the short reminder and the subtask. However, regarding the long reminder the results remain unclear. Figure 5.14 shows the number of test persons in Group D that allowed the interruptions. The fact that each subject who tested the *negotiation* strategy rejected at least one of the interruptions is also of interest in this context. We have observed similar user reactions regarding all strategies in case the user has been asked if the main task could be interrupted: one third of the subjects denied. The most common explanation was the desire to focus on the main task.

5.3.3 Conclusion

The main goal of our work was to investigate an efficient way of switching topics in order to improve the user acceptance and the SDS performance within real-life situations. For this purpose, we have examined four different methods for handling interruptions. According to the represented evaluation results all of them demonstrate strengths and weaknesses. The *explanation* and the *negotiation* strategy showed high scores in terms of reliability and a clear separation to the main task.

It seems that the *explanation* and the *negotiation* strategy also support the user to memorise the reminders. However, the *baseline* and the *discourse* strategies showed the advantage of being less irritating. Within real-life situations, when the interruptions are realistic and, for example, critical, the differences regarding the degree to what the information is perceived may decrease.

Nevertheless, an important—and unexpected—benefit of the *explanation* and the *negotiation* strategies was the impact they had on the overall dialogue quality. Groups C and D received a significant lower number of non-understandings than the other two groups. Furthermore, the total dialogue duration of Groups C and D was significantly lower as well. Our hypothesis that topic switching strategies may compensate the drawbacks we have faced during the initial system test seems to be a correct one. Overall, the *explanation* strategy turns out to be the most promising approach. Only the *negotiation* strategy can compete—however, since this strategy allows to totally suppress interruptions, it is only partly useful within a realistic scenario. In the following section we present the results of the second evaluation regarding dialogue optimisation: a comparison of repair strategies that have been applied to the ASDM.

5.4 Repair Strategies

In this section we focus on the results of an evaluation we have carried out to investigate the benefits and drawbacks of different repair strategies that can be applied to the VoiceXML-based output of the OwlSpeak ASDM. In Sect. 3.6.2.2 we have introduced three strategies we utilised to assist the user in case the ASDM and/or the user detects that an error has occurred during the dialogue evolved. We have analysed the following strategies:

- The *Modified MoveOn* strategy may be applied if the user input is interrelated. The hypothesis is that it may be more effective to skip individual confirmations for a couple of consecutive questions. This would lead to the availability of more information when making a combined confirmation. Such an explicit confirmation may serve as a starting point for automatically correcting the user input by, for example, replacing the item of information with the lowest probability.
- The *Reprompt* strategy implicitly confirms any user input. In case of a misunderstanding the user must explicitly interfere with signal words such as "wrong" or "false". This triggers the system to inform the user that the desired correction will be attempted, followed by a question reprompt.
- The *Help* strategy works similarly as the reprompt strategy, except that a help dialogue, explaining the requested information, is added to the question reprompt.

In general, our prototype can automatically apply all strategies to a specific dialogue. Only the *Modified MoveOn* requires additional integration effort for a generic usage of the strategy. Both the *Reprompt* and the *Help* strategy only require additional information that may be provided by the dialogue models (e.g. grammars to match "wrong" utterances and help requests). For this investigation we refrained from implementing all strategies to run with the ASDM but only altered the VoiceXML output. Hence, we did not implement the strategies to be applied dynamically to the dialogues but hardcoded the different dialogues together with the specific strategy. In the following we discuss the experimental set-up that has been used and present details on the implementation of the evaluation.

5.4.1 Experimental Set-up

We decided to establish a straightforward experimental set-up for investigating the repair strategies. In order to allow the users to rate the dialogues as unbiased as possible, we tried to minimise the requirements and the cognitive load of the users. Three comparable dialogues each combined with the three different strategies were designed. The dialogues were implemented using the OwlSpeak prototype that has been used for the initial system test without any enhancements (see Sect. 5.1). For the evaluation we applied the online available commercial voice platform TellMe.Studio.[2] It provides VoiceXML interpreter, telephony gateway, TTS, and ASR. Using English as main dialogue language influenced the selection of subjects.

Users called the system via Skype using a fixed hardware set-up. In this study we do not focus on ASR. For this reason we carried out the tests under optimal conditions (no noise etc.) using a high-quality headset. To be able to abort the dialogue in case of an issue that cannot be solved by the subject, the supervising person monitored the dialogues. However, this was usually not necessary and most of the time, the reason for an interruption could easily be resolved. In case of software or hardware issues, which were not caused by the user or by the SDS, the affected part of the experiment was repeated. In total, 23 test persons evaluated the different strategies. We divided the subjects into two groups of equal size: one group consisted of the "computer experts" (e.g. engineers or computer scientists) and the other one consisted of users without any technical background. A requirement for the test persons was to be proficient in English. In the following section we discuss how the dialogues we used for evaluating the strategies were designed.

5.4.1.1 Test Dialogues

Each subject was asked to test the repair strategies. To facilitate the comparison, the three strategies were implemented using different scenarios. We designed a dialogue concerning a recipe selection, a dialogue for cinema ticket reservation, and a dialogue regarding ski rental. For each dialogue we deployed three versions,

[2]https://studio.tellme.com/

one for each strategy. Thus, in total we developed a set of nine dialogues. In order to guarantee a fair competition and, most importantly, not to favour one particular strategy in advance, we established the following principles:

- All dialogues for all scenarios must be equal regarding usability and user-friendliness.
- Each subject should at least evaluate six dialogues with different repair strategies.
- The dialogues must be evaluated using random sequences of the scenarios.

To keep the complexity of the dialogues low, they are as rudimentary as possible but—in principle—still useful in practice. Each dialogue consists of three questions regarding the different topics to be solved followed by a summary question to confirm the user input. The possible answers of the subjects are also kept as short as possible, at best a single word. The combination of a simplistic grammar with the optimal conditions resulted in a high ASR performance. Thus, we assume that recognition errors do not influence the outcome of the evaluation. Since we did not want to rely on *unexpected* system failures, we simulated the misunderstandings in order to repeatedly apply the repair strategies within the same context. This solution allowed for an exact simulation of the dialogue situation we intended to investigate. An exemplary dialogue is presented in Table 3.9 (Sect. 3.6.2.2). This dialogue shows how we falsified the ingredient that actually has been correctly recognised. In this example we demonstrate the *Modified MoveOn* strategy with a correct "guess" that repairs the previously (faked) misunderstanding.

Tests of the experimental set-up showed that a number of six dialogues for each subject is ideal. The maximum number of nine dialogues (three domains each combined with one of the three strategies) would have bored the volunteer subjects and therefore would have biased the results. We prescribed random sequences of six dialogues, so that each strategy at least one time was used as the first dialogue. This dialogue was usually the one where the participants had most problems with. During the following dialogues they got used to the system because of the similar design of the dialogues. To decrease the issues especially in the beginning, the users received a short introduction providing the required information for successfully concluding the dialogues.

5.4.1.2 Questionnaire

After completing a dialogue, the subjects were asked to fill in a questionnaire. As with all evaluations we adopted SASSI to establish the survey. The comprehensive questionnaire has been shortened to 17 questions. To keep the subjects motivated, we limited the duration of the experiment to 30 min. In order to analyse the results we merged the user rating regarding different questions into generalised factors. For the evaluation we established the following five quality factors:

EFF *Efficiency* describes how efficient the dialogue is. How well can the dialogue flow be comprehended by the user?

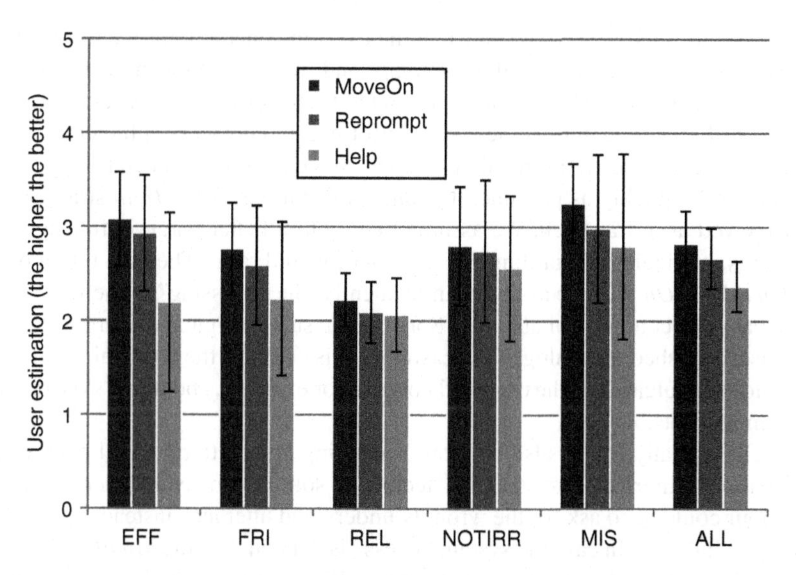

Fig. 5.15 The subjective user rating of all users on average. The error indicators show the standard deviation

FRI *Friendliness* describes how user-friendly the system is. Was it pleasant to use the system?

REL *Reliability* describes how reliable the system is. Does the system cause many mistakes and does the user have any issues in understanding it?

NOTIRR *Not Irritating* describes how the user been irritated by the SDS. Did the system work like expected and foreseen? A low rating of this measure means that the user was confused by the system.

MIS *Mistakes* describes if it is clear for the user how to react in the case of a misunderstanding. Furthermore, it describes if it was easy for the user to correct the issue.

ALL *Overall* reflects an overall subjective estimation by the user of the entire SDS. In the following this metric is calculated from the average of EFF, FRI, REL, NOTIRR, and MIS.

The questionnaire is presented in Appendix E (Figs. E.1–E.3). In the following section we present the results of our investigation and analyse the collected data.

5.4.2 Evaluation Results

We divided the test field into two groups: *beginners* and *experts*. In the following we analyse the complete test field before taking a look at the results of the two specific groups. Figure 5.15 plots the overall results of the subjective rating.

The columns indicate a tendency that the users in general preferred the *Modified MoveOn* strategy, followed by the *Reprompt* and, with a significant distinction, by the *Help* strategy. The *Help* strategy was rated to perform worst. A reason for this may be the barge-in feature of the voice platform. It did not work sufficiently, which forced us to deactivate it during all dialogues. Since the users wanted to execute the corrections as quickly as possible, the long explanations of the *Help* strategy may be perceived as a drawback. We assume that, with a better barge-in functionality, the scores, particularly regarding EFF and IRR, would raise. The high rating of the *Modified MoveOn* must be assessed ambivalently: the "guessing" of the wrong user input was correct in 90% of all cases. Some of the subjects ignored a wrong "guess" and falsely finished the dialogue. A reason for this could be the artificial situation of the dialogue evaluation—the users did not talk, for example, about *real* skis, recipes, or cinema tickets.

Psychologically it may also be more frustrating having to do "it all over again" than simply ignoring a mistake. A technical solution to avoid such a system behaviour could be to ask for the wrongly understood utterance instead of repeating the entire dialogue. In case the system "guess" is dependable, the *Modified MoveOn* strategy bears the advantage to cloak errors. If an error occurs in the beginning, the user may not notice it. Thus, he stays positive and motivated. Furthermore, the dialogue can be continued more smoothly and chances that the system can (self-)resolve a contradiction increase. Hence, if the underlying mechanisms work sufficiently, our results indicate that this is the strategy the users are most satisfied with. Since all strategies were tested by all subjects it was possible to apply a *paired* Student's t-test (Mankiewicz 2000) so as to test the specific results for statistical significance. The analysis of the EFF metric reveals that the subjects rate the *MoveOn* and the *Reprompt* equally but compared to the *Help* strategy EFF is rated significantly lower (P value = 0.001).

We received similar results regarding the FRI metric: the subjects rate the *MoveOn* and the *Reprompt* similarly but if we pair-wise compare *MoveOn* & *Help* with *Reprompt* & *Help*, the Student's t-test calculates a P value of 0.001 and 0.019, respectively. Again, this underpins the low rating of the performance of the *Help* strategy. Regarding the REL and the NOTIRR metric, the results do not show a significant tendency of the subjects' rating: all approaches were rated relatively low. The comparison of the results of the MIS metric also reveal a significant difference between the rating of the *MoveOn* and of the *Help* strategy (P value = 0.033). However, both pairs *MoveOn* & *Reprompt* and *Reprompt* & *Help* do not appear to be rated significantly differently. As mentioned above, the ALL metric is calculated from the other metrics and presents the results only by way of illustration. Thus, these results cannot be tested for significance. Besides this, we experienced major differences regarding the behaviour during the evaluation and the handling of the dialogue between subjects who were experienced with computers and persons without technical background. The experienced users (shown in Fig. 5.16) encountered significant fewer issues with the dialogues and successfully finished them most of the time.

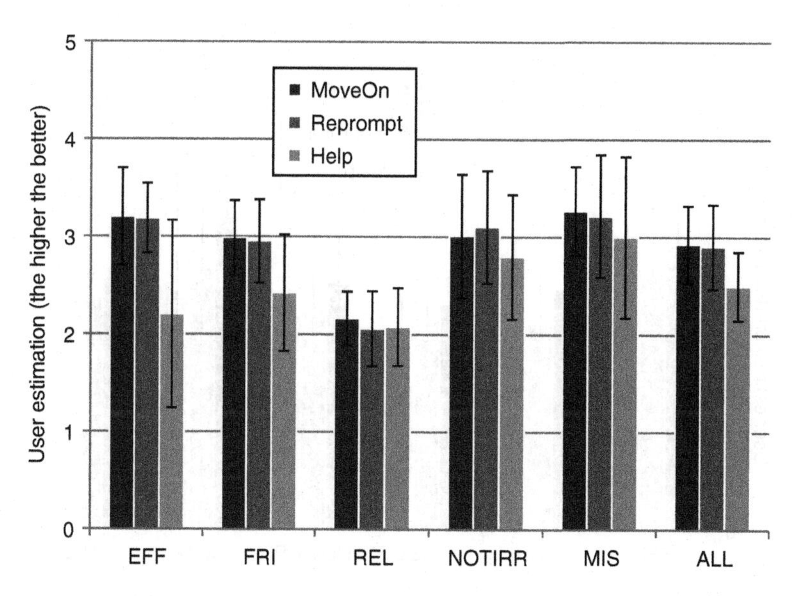

Fig. 5.16 The subjective user rating of the experienced users on average. The error indicators show the standard deviation

Once they got used to the voice interface, the experts also got less patient and therefore especially appreciated the efficiency of the *Modified MoveOn* strategy. They got annoyed by the *Help* strategy due to its lengthy. However, if they encounter a serious issue, they were happy about a detailed help. The experts' rating regarding the EFF and the FRI metric are equal to the results presented above. EFF and FRI are rated almost identically concerning the *MoveOn* and the *Reprompt*. However, again the *Help* strategy is rated to perform significantly worse regarding the two quality factors (P values ≤ 0.02). Compared to the high ratings of the *MoveOn* and the *Reprompt* strategies, the REL metric is rated significantly lower concerning all strategies. Since we *falsified* an erroneous system behaviour this result was expected. On average, the experts rated the *MoveOn* and the *Reprompt* strategies slightly better than the *Help* strategy. Regarding these factors, however, the differences are not significant.

Figure 5.17 plots the rating of the unexperienced users. A few subjects had a general disapproval of talking to a machine and therefore experienced issues with the different dialogues. Additionally, the SDS confirmed their attitudes since all users experienced at least one—the *falsified*—mistake. The greater diversity of answers of these subjects can be derived from the higher standard deviations of the EFF, the FRI, the NOTIRR, and the MIS scores in Fig. 5.17 compared to the standard deviations in Fig. 5.16.

The Student's t-test calculates the same significant differences between the pairs *MoveOn* & *Help* (P value = 0.015) and *Reprompt* & *Help* (P value = 0.038) regarding the EFF metric. The *MoveOn* strategy, however, was rated to be significantly more reliable (REL) than the *Help* strategy. On average, the *Reprompt*

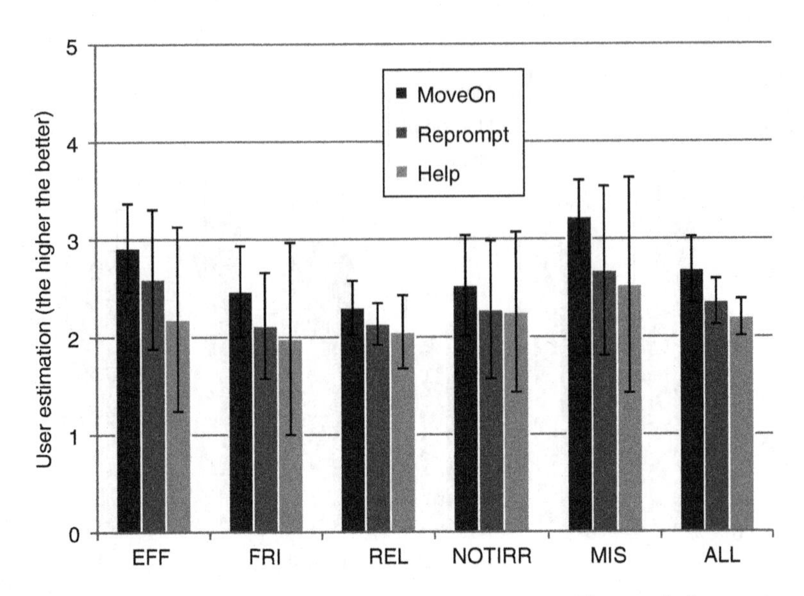

Fig. 5.17 The subjective user rating of the non-experts on average. The error indicators show the standard deviation

strategy is rated significantly lower by the unexperienced users than by the experts (*P* value = 0.04). A similar result is revealed by the FRI scores: the unexperienced users voted the three strategies equally. The FRI score of the *Reprompt* strategy is rated significantly lower than the same score rated by the experts (*P* value = 0.001). In total, all strategies regarding nearly all quality factors were rated lower by the unexperienced users. Concerning the NOTIRR quality factor, we did not observe any significant different rating by the non-experts. As expected, the expert users rated all strategies better—they are more used to computer interfaces in general. The *MoveOn* strategy shows a tendency to be rated better than the *Help* strategy by the non-experts (*P* value = 0.077). Figure 5.17 depicts an average score of 3.3 with a low standard deviation. In relation to this, the *Help* strategy is rated below 2.5 with a high standard deviation. Figure 5.18 highlights the differences between the ratings of the two user groups.

On average, the *Reprompt* strategy showed the highest difference when we compare the experts with the unexperienced users: regarding EFF, FRI, and NOTIRR, the experts rated this strategy to be significantly better. The REL factor is rated equally by both groups. The experienced user honoured the naturalness and the intelligence of the *Modified MoveOn* strategy. It is also of interest that the non-experts were more irritated by the surprising "guess" of this dialogue mechanism. For this reason, we assume that it is not beneficial to confront users with strategies they are not able to comprehend. The experts had less problems in correcting their input using the *Reprompt* and the *Help* strategy. This is probably a reason for the significant different rating of FRI. A general observation we made was that women

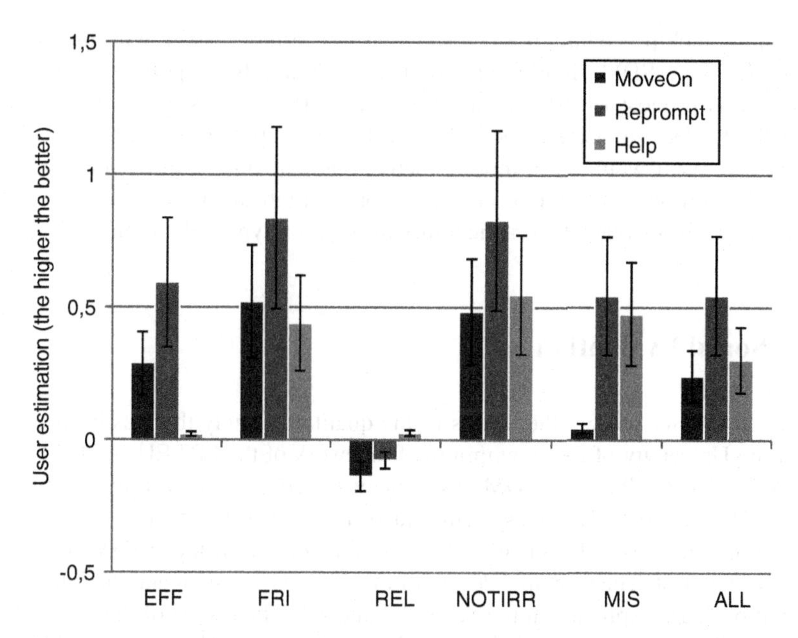

Fig. 5.18 The difference between the subjective rating of the experienced users and of the beginners. A value around zero means that both groups agree. A positive value indicates a better rating by the experts. The error indicators show the standard deviation

were less bothered and more patient by the detailed help prompts than male subjects. We argue that especially when dealing with novice users the perception of repair strategies significantly differs. Therefore, user profiles would be required in order to select the most appropriate strategy on a per-user basis.

5.4.3 Conclusion

We propose to incorporate user characteristics when choosing a recovery or repair strategy. Factors such as technical background, mood, and ability of handling the SDS influence a successful and smooth dialogue flow. Therefore, we aim at a system that maximises the user satisfaction would also increase the task success rate. In general, it should be the purpose of the ASDM to integrate background information about the user into the dialogue generation. The system must be aware of the user skills, for example, to adapt the repair strategies accordingly. Prosody features and other factors should be comprised to determine if the chosen strategy was the correct one and which ones should be selected subsequently. For example, if the user seems to be annoyed of the repeated and detailed help, it is the duty of the ASDM to react accordingly. For this evaluation we distinguished between two groups of users making use of our strategies in a different manner. While experts preferred a rather direct and quick procedure, the novice subjects had more issues with the system. Therefore, the ASDM should support them accordingly. A policy for recovery strategies as proposed in Bohus and Rudnicky (2005) seems to be promising. If the ASDM selects the strategies with regard to the expected user satisfaction, we believe the result would be a more effective dialogue. Our results can help here, since they show how different user groups cope with the analysed strategies. In the following we investigate the practicability of our prototype and our methods.

5.5 Social Evaluation

In this section we present the results of the qualitative study that has been carried out at the University of Essex within the framework of the FP7 EU-funded Project ATRACO. The OwlSpeak ASDM has been evaluated as a full-working component integrated into the ATRACO system and installed in the iSpace. For the social evaluation we utilised the same SDS in combination with the ASDM as for the initial system test. The principle design of the ATRACO system and the integration of OwlSpeak are explained in the Sects. 2.2 and 4.1. The results of the entire study covering all components have been published as part of van Helvert et al. (2011). In this section we focus on the results that have been collected regarding the ASDM. In the following we present the study outline before we discuss the qualitative results that show a clear tendency towards a learning effect, the test persons experienced. In the second part of the section we compare this results with the quantitative data that has been protocolled and finally discuss the findings in Sect. 5.5.4.

5.5.1 Study Outline

The most important difference of this study compared to the other experiments and evaluation sessions we have conducted is that the ATRACO prototype supports a "free-play" mode. This means that the subjects were not bound to specific tasks or guidelines but interacted with the system within a particular Activity Sphere in an unconstrained way. The format of the study was focussed on three Activity Spheres for different purposes (see Sect. 2.2). During the integration of the prototype three scenarios have been realised:

AS1: Entertainment The Entertainment Sphere is centred on the idea of relaxing in the living room. Involved components are a lighting control, a heating control, a music player, and a photo viewer. All components are accessible via a visual interface (on an iPad, the main TV and a Laptop) and via the ASDM providing the adaptive voice interface.

AS2: Work The Work Sphere integrated environmental controls with home banking, document production, and other forms of work and study option. Involved

Table 5.3 The format of the iSpace evaluation (cf. van Helvert et al. 2011)

Schedule	Action
Prior to the initial evaluation session	Present an overview of the prototype and the broader ATRACO concepts, allowing participants to question and clarify their understanding. Introduce the "gap" concept with examples (Dervin et al. 2003). Talk to the participant about being conscious/mindful of his own gaps as they move through the experience and encourage them to raise their hand each time a gap is encountered
For each AS	Run free play session and record on video. Review the video with the participant and conduct the interview
At the end of the final session	Conduct a closing interview with participants who have participated in two or more sessions

components are a lighting control, a heating control, a music player, links to home banking, calendar, news, and the social network Facebook. Furthermore, Google docs and a spread sheet application were integrated. All components and services were accessible via a visual interface (on the workstation screen) and via the ASDM providing the adaptive voice interface.

AS3: Sleep The Sleep Sphere is centred on patterns of activity in the bedroom at the end of the day. Involved components are the light control, the heating control, the music player, a photo viewer, and a security function that allowed securing the bedroom. Again all components were accessible via a visual interface (on an iPad, the main TV, and a Laptop) and via the ASDM providing the adaptive voice interface.

In Appendix A we present the dialogues that were implemented and tested as part of this evaluation. Each AS was individually tested by all test persons during a dedicated session. Thus, each subject had the chance to become increasingly familiar with the system over three separate sessions. It was planned to interview the participants after each session. After the final session a closing interview was conducted to gather overall impressions.

Table 5.3 shows an overview of the evaluation format. In total, ten test persons have participated in the evaluation. Only six of them, however, attended all three AS-specific sessions (four male and two female subjects). In the following we focus on these subjects since they had the chance to move from a superficial to a deeper understanding of the main ATRACO concepts and the corresponding interfaces. In order to differentiate the subjects, we use IDs from P1 to P6 both for the comments and for the quantitative analysis. In the following section we present the most important comments, the subjects made regarding the voice interface, i.e. the OwlSpeak ASDM.

5.5.2 *Qualitative Results*

All user comments we present in this section are taken from the interviews that were conducted for each AS. The comments together with an in-depth analysis and the outcome of all interviews are published as part of the ATRACO project deliverable D27 (van Helvert et al. 2011). The comments are presented sphere-wise starting from AS1 via AS3 to AS2. This means that the subjects were not familiar with the system in the beginning. We will see that for AS3 and especially for AS2 the comments regarding the voice interface significantly change. During the first evaluation session we received comments regarding the high number of repetitions that had to be uttered, regarding the raised voice and the curt nature of the commands, and regarding the general idea of a voice interface that works efficiently. Regarding the high number of repetitions that had to be uttered, i.e. the number of non-understandings, we received two comments that are of interest. Subject P2 stated the following:

> "Even when it worked, it responded slowly. I wasn't sure if it received the message so I had to say it three or four times... I know it's an interface to the system but you start thinking that it's something wrong with you... you are thinking it should work but it's not working so maybe it's how you pronounce it or something... I don't think I'll be using it."

The SDS did not catch the input of P2 sufficiently. This caused severe issues. The user also indicates that he started thinking that these issues were caused by himself. This is very problematic since the non-understandings are, of course, a technical issue that relates, for example, to the set-up of the microphones. The second comment by subject P3 points into a similar direction:

> "Irritating because it's like 'do it' and it doesn't do it. I got more angry if I had to repeat it. I prefer to do it myself instead of repeating it... it's like talking to a wall."

Again, the high number of non-understandings is emphasised by this subject. A further drawback that was criticised by subject P4 relates to the wording of the commands and their curt nature:

> "And if I would talk louder than I did, somehow it would not feel natural."

The main issue was that the phrases chosen for the commands were not natural and comfortable for the subjects. This was contradicting to the requirements we identified during design time. We assumed that short and simple commands are easier to use and easier to keep in mind. However, the subjects have also been asked how they think about a voice interface with a higher recognition rate. Subject P1 and P6 agree that a *working* system would definitely be beneficial.

> "If you imagine something like Jarvis from Iron Man then yes! If it was a lot more responsive then I'd probably go for it."

> "That sounds to me that it would be really, really good... I like the idea of sitting on the settee saying 'TV on, music off, heating up', that's the attractive thing about it."

As mentioned, the subjects were not familiar with the system during this first evaluation session. During the second session where the subjects used AS3 the overall feedback was slightly better. Subject P1 stated the following:

> "That was the first time I was able to use it and it worked pretty well."

This tendency could be continued during the course of the third evaluation session. The users get used to system and, besides the fact that they still experienced some unsuccessful commands, their enthusiasm to use the voice interface increased. The subjects P2, P5, and P6 agreed that a voice interface at home would be useful and beneficial.

> "It would really be a good thing—only if you are alone though, because if there was a noise or something it would be useless."

> "I think the ability to operate different things by voice command is good—when it works [laughs]... Once you've got used to it, I think it could be very useful."

> "I just want to say that with the voice control, I think it would be wonderful, certainly for an office at home."

After the last session where the subjects used AS2, test person P4 even summarised our observation of a learning process and detailed how he got used to the interface during the three sessions:

> "I also had a feeling that, even when I wasn't successful with the voice commands, that it was more comfortable. I was more comfortable with it... [not responding] than the previous two times. Because I felt very weird the first time and the second time it was still uncomfortable when it didn't respond... I've probably become more used to the idea that I can use my voice to do something or to give a command... The first time it was more probably the inhibition more than frustration. The second time it was more the frustration—'why doesn't it work!'... No I think that would be alright, yeah—but it should work more often than not!"

5.5.3 Quantitative Results

In the previous section we presented our observation showing that the subjects went through a learning process. At the end of the study, this process led to a much better subjective rating than after the first session. In this section we present the objective results, we collected during the social evaluation. We assume that the objective data underpins the subjective user estimations and the observed learning process. Figure 5.19 shows the number of non-understandings per subject that occurred during the 18 evaluation sessions compared to the number of spoken commands the system understood correctly and the unrelated utterances.

Each session took between 22 and 60 min. In order to render the numbers comparable we have normalised them to a duration of 30 min. On average, we received a high number of non-understandings. In total 212 non-understandings and 252 understandings were recorded. Furthermore, the system detected 162 unrelated

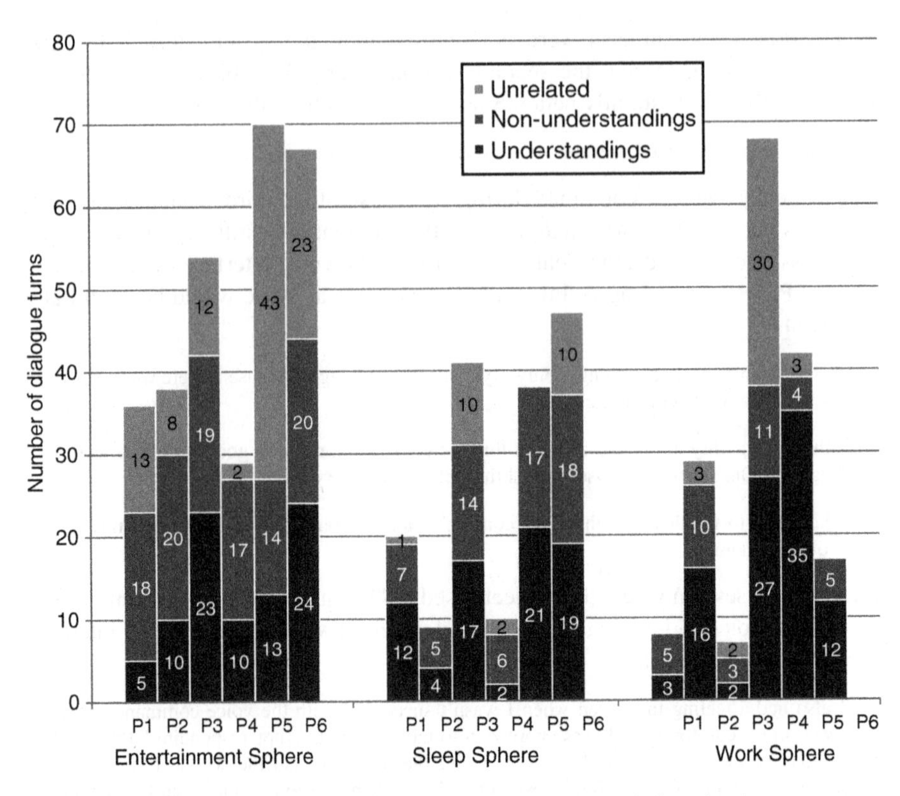

Fig. 5.19 The number of understandings, of non-understandings, and of unrelated user utterances

utterances that the system rejected correctly. Even though these numbers are quite sobering, it must be noted that the ASDM is robust against situations where the user is not directly talking to the system. Critical situations are when user is talking to a third person, when he is just interjecting utterances such as "Oh, my god", and when he uses fill words that explicitly should not be recognised. Figure 5.20 shows the average percentage of correctly understood utterances, non-understandings, and unrelated phrases. Concerning the Entertainment Sphere, the objective results approve the subjective user comments. On average, the users received nearly one-third more non-understandings than understandings, which reasonably is an issue.

The ratio between understandings and non-understandings increased during the Sleep Sphere. However, the subjects were still confronted with 41% of falsely rejected input. Due to the low number of test persons the differences between the Entertainment and the Sleep Sphere are not significant. A comparison between the Sleep and the Work Sphere reveals a similar result regarding the average percentage of understandings: the system was able to understand twice as much utterances correctly than during the first session. Furthermore, the subjects were confronted

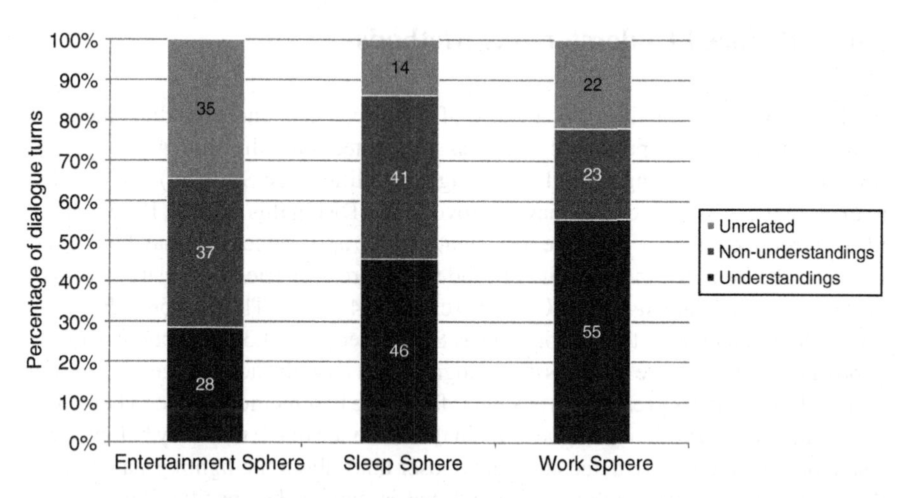

Fig. 5.20 The average percentage of understandings compared to the percentage of non-understandings and the unrelated user utterances that have been recorded during 18 evaluation sessions

with fewer non-understandings. The improvement from the first session to the last session is significant: the variance of analysis for multiple comparisons (Stoline 1981) calculates a P value of 0.002. This result correlates with the subjective user estimations that have been collected during the interviews. In the following we conclude this section and discuss the results.

5.5.4 Conclusion

The evaluation shows that after three sessions, the users where satisfied with the performance of the SDS. However, this satisfaction strongly depends on the individual learning stage regarding the control of the voice interface. The investigation reveals that even a simplistic command-and-control voice interface is not as intuitive for the users as, for example, a GUI for the same purpose would be. The results motivated us to investigate the options we have to handle non-understandings more intelligently. Furthermore, we implemented several methods that can be utilised, combined or separately, to improve the performance of the ASDM. The outcome of the social evaluation demonstrated the importance of *intuitive* voice interfaces. Especially, for model-based ASDM we see several options to improve the intuitivity of the dialogue. The technical realisation of these approaches has been presented in Sect. 4.2.3.4. In the following we present the results of a comparison of these novel approaches that could lead to fewer non-understandings without increasing the number of misunderstandings.

5.6 Advanced Understanding Methods

In this section we analyse the results of the advanced understanding methods evaluation. After the prototype that was integrated into the ATRACO system, revealed a lack regarding its understanding capabilities (see Sect. 5.5), we decided to explore the possibilities that may improve the ASDM in this respect. Based on the ideas introduced in Sect. 3.6.3 we investigated three approaches that may be utilised to render the interface more intuitively and thus more user-friendly: *keyword-based*, *blurred-keyword*, and *semantic-keyword* (cf. Sect. 4.2.3.4). These approaches base on the dialogue models that are defined as SDOs (see Sect. 4.3). The purpose of the approaches is to provide a successful recognition in case the actually generated (but in its individual parts prescribed) grammar fails. The first method utilises a keyword-spotter to detect to which dialogue domain the utterance of the user may belong to. If a suitable domain can be detected (e.g. the user talks about the lighting), a keyword that relates to a corresponding command that is part of the specific domain must be spotted, subsequently. In case the system detects such a keyword (e.g. the word "low" is part of the utterance), the ASDM automatically generates a confirmation dialogue to prove its assumption.

The second approach combines the method of keyword-spotting with free-text recognition and a Levenshtein-based comparison of the input and the according keywords. We assume the *blurred-keyword* comparison of the keywords defined as part of the dialogue model with the most probable results of the recogniser to be more robust against recognition errors. Furthermore, the comparison may avoid issues caused by homonyms and homophones. In German, most homophones can be covered under a Levenshtein distance of *one* (e.g. "Mainz" and "meins"). Therefore, if the recogniser provides the word "meins", a possible match with a dialogue domain that provides "Mainz" as keyword would be possible. The (*semantic-keyword*) approach integrates a semantic knowledgebase that allows to find semantic similarities between the keywords and the recognition results. Especially to facilitate the understanding and the comprehension of intuitive user utterances, our approach seems to be promising. In the following we describe the experimental set-up before we analyse the results of our experiments.

5.6.1 Experimental Set-up

We decided to implement a scenario related to home automation for this evaluation. We intended to facilitate the comparison to the results obtained during the social evaluation (see Sect. 5.5). Again, we defined several dialogue domains for the devices that were automatically combined into a coherent dialogue. The generated command-and-control dialogues are similar to the ATRACO dialogues (cf. Appendix A). Figure 5.21 shows the virtual room that provided a visual feedback to the commands the subjects were asked to utter intuitively. Since fostering the intuitiveness of the OwlSpeak-based SDS was a main motivation of our approach,

Fig. 5.21 Virtual room used to visualise the test bed

we did not reveal the subjects, the commands, the SDS was actually able to understand. Thereby, all subjects were forced to control the virtual environment as they personally supposed. The OwlSpeak system delivered minimal grammars for the six devices that were integrated within the virtual environment. In case a user is aware of the possible commands, the SDS works appropriately with these grammars. This however, is not what we intended to evaluate. Instead, the main aim was to evaluate how users *intuitively* cope with such a system.

As depicted in Fig. 5.21, a ventilator, a TV, a stereo, the heating, the sun-blinds, and a lamp could be controlled using commands such as "switch the light on" and "volume up". These commands were part of the original SDOs. For the evaluation we enriched these ontologies with keywords for all domains and for their corresponding commands. For the music domain, we introduced keywords such as "audio equipment", "hi-fi", and "music". Table 5.4 lists the keywords we have defined to detect the domains and the commands in German language. A benefit of the compact and minimal grammars is that they clearly define a specific command. This facilitates the ambiguity detection and requires limited maintenance effort. Minimal grammars are not as prone to recognition errors as complex grammars that may overlap with other grammars are. Minimal grammars, however, do not support intuitive communication. The virtual room provided the visual feedback encoded by using the colours red, blue, and green. Deactivated devices are coloured red and activated devices are coloured green. Blue devices are currently changing their state, e.g. the volume of the device changes. The explanation of the virtual room was part of the short introduction, the subjects received.

Table 5.4 The domain-keywords, commands, grammars, and command-keywords of the six dialogue domains that were involved in the evaluation

SDO	Domain-Keywords	Commands	Grammars	Command-Keywords
light	Licht, Lampe, Leuchte	LightsOff_move	(licht aus)	aus (off)
		LightsOn_move	(licht an)	an, ein (on)
		high_move	(helles licht)	heller, hell (bright)
		low_move	(mattes licht)	dunkler, dunkel (dark)
heating	Heizkörper, Heizung, Heizgerät	higher_move	(heizung wärmer)	warm, wärmer (warmer)
		lower_move	(heizung kälter)	kalt, kälter (colder)
		off_move	(heizung aus)	aus (off)
		on_move	(heizung an)	an, ein (on)
music	Stereoanlage, Musik, Audio, Radio	exit_move	(musik beenden)	ende, stop, aus (off)
		play_move	(musik starten)	abspielen, an (on)
		volUp_move	(musik lauter)	laut (volume up)
		volDown_move	(musik leiser)	leise (volume down)
tv	TV, Fernseher	off_move	(fernseher aus)	aus (off)
		on_move	[(fernseher an) (fernseher ein)]	an, ein (on)
		loud_move	(fernseher lauter)	laut (volume up)
		low_move	(fernseher leiser)	leise (volume down)
sun-blind	Jalousie, Rolladen, Rollo, Sonnenschutz	close_move	[(rolladen schließen) (jalousie zu)]	zu, runter (up)
		open_move	[(rolladen oeffnen) (jalousie oeffnen) (rolladen auf)]	auf, hoch (down)
blower	Ventilator, Lüfter	fanOff_move	(heizung wärmer)	aus (off)
		fanOn_move	[(ventilator an) (ventilator ein)]	ein, an (on)

The subjects' goal was simple and comprehensive in unison: they were asked to control the environment. Forty subjects participated in the evaluation: 30 male and 10 female users. All test persons were at least fluent German speakers. Only three subjects were non-natives but were proficient in German. For the evaluation, we used a similar set-up as for the evaluation of the topic switching strategies. However, we integrated a second recogniser (the Microsoft Speech API) to perform a subsequent grammar-based or OOV recognition in case a non-understanding occurs. The SDS that was used consists of the following modules:

- Spoken Dialogue Manager: OwlSpeak ASDM
- Speech Platform: Voxeo Prophecy 10
- MRCP-Server: Loquendo Speech Suite 7 (including TTS and ASR)
- SIP-Client: Linphone
- External Recogniser: Microsoft Speech API (grammar-based + OOV)

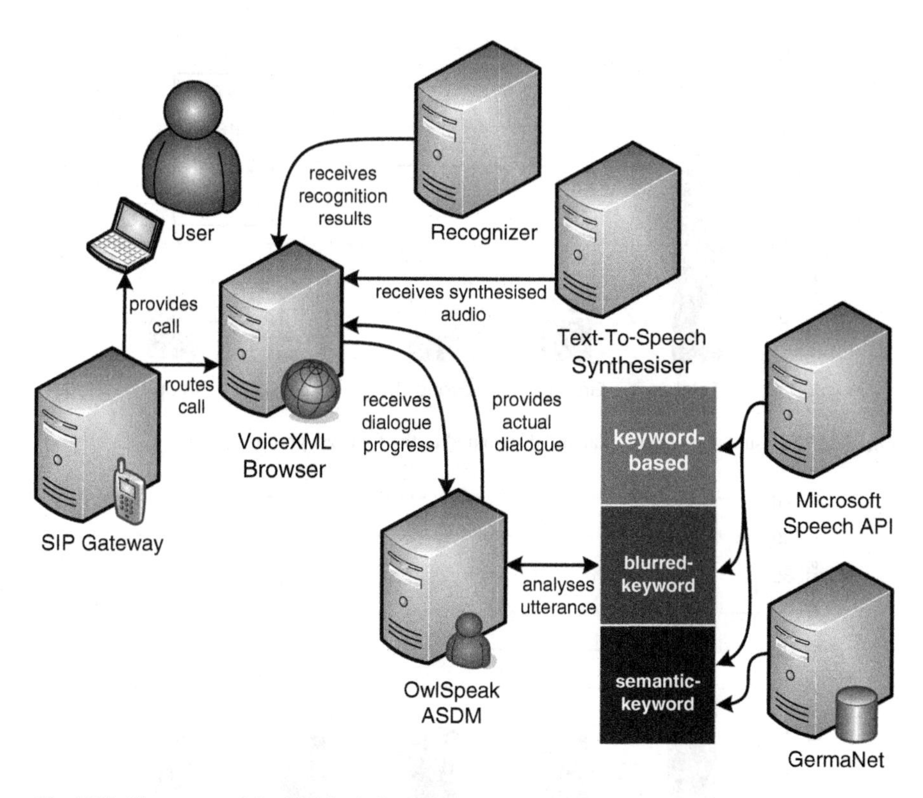

Fig. 5.22 The set-up of the SDS including the three approaches, the MS SAPI, and GermaNet

Figure 5.22 illustrates an overview of the SDS. The main part was integrated similarly to the system that has been used for the evaluations described in Sect. 5.1 and in Sect. 5.3. The system set-up used for the actual experiment, however, incorporates the three understanding approaches, the MS SAPI as external recogniser, and GermaNet as semantic knowledgebase. If a non-understanding occurs, the OwlSpeak ASDM triggers the specific understanding enhancement. Afterwards, in case of a positive detection, the system initialises a confirmation dialogue. In order to allow for a comparison of the three approaches with the baseline system, we divided the 40 subjects into four groups each consisting of 10 people. Group A used the baseline system, Group B the keyword-based approach, Group C the blurred-keyword approach, and Group D the semantic-keyword system.

Figure 5.23 depicts the subjects' age. A majority of the users were aged between 18 and 40 years while nearly all subjects were younger than 60 years. On average, the subjects of Group A and C are 10 years younger than the Group B and D users. The differences, however, are not significant and therefore we estimate that the distribution of the subjects amongst the groups is homogenous regarding the age.

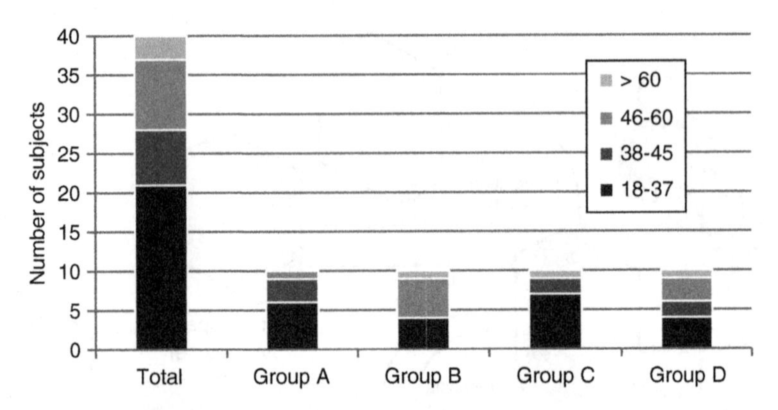

Fig. 5.23 The subjects' age in total and distributed to the four test groups

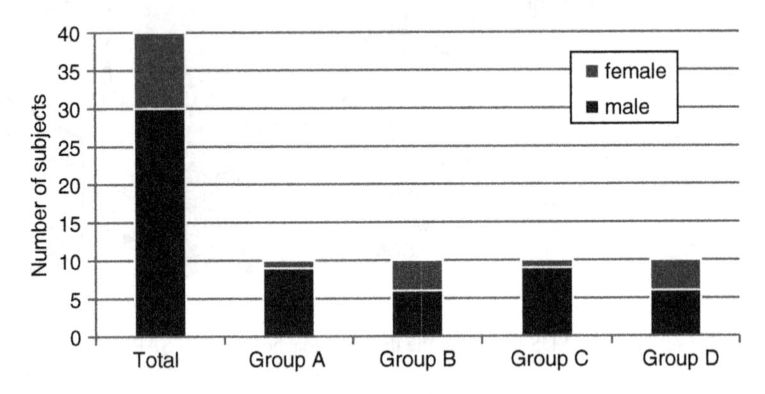

Fig. 5.24 The distribution of the subjects' gender in total and group-specific

Figure 5.24 plots the number of female and the number of male subjects. Seventy-five per cent male subjects and 25% female subjects participated in the evaluation. Group A and Group C consisted of nine male users and one female subject. For Groups B and D we achieved a more equal ratio: each group consisted of six male and four female subjects. Before we started the actual test run, we asked the subjects for their expertise regarding computers, IT in general, and SDSs.

Figure 5.25 plots the users' estimation of their technical experience in total and itemised per group. The scale of the figures maps the users' ratings (one = worst and seven = best) to different colours. Dark colours represent novices and bright colour highlight experts. Figure 5.25a shows that a majority of the users were experienced with IT in general. Group A and C are more skilled regarding computers than Group B and D are. In total, the subjects were less experienced regarding SDS (see Fig. 5.25b): 25 subjects rated their experience below neutral. The distribution of SDS novices and experts amongst the four groups looks similar to the arrangement regarding IT and computer experience. Group A and Group C consisted of less novices than Group B and Group D.

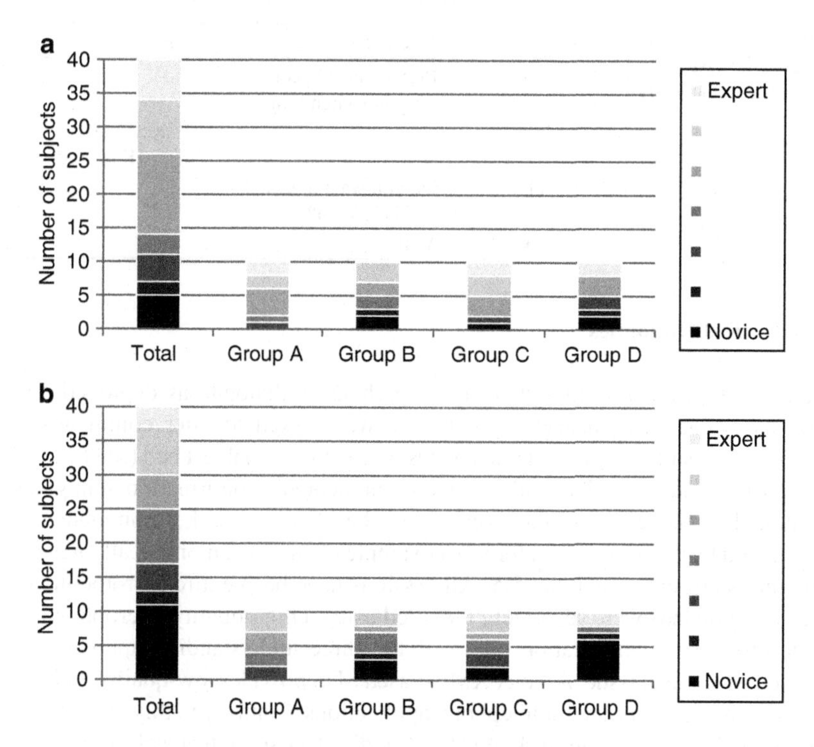

Fig. 5.25 The expertise of the subjects regarding IT and SDS. The scale ranges from the darkest colour (novice) to the brightest colour (expert). (**a**) The subjects' expertise regarding computers and IT in general (**b**) The expertise of the subjects regarding Spoken Dialogue Systems

More than half of the Group D subjects indicated that they do not have any experience with SDSs at all. Overall, Group A—the baseline group—consisted of less novices than the other groups. This distribution was desired, since the baseline system was not designed to be effectively used by novice users. Since the other groups took advantage of the understanding enhancements, experts did not influence the results unduly. Each group was asked to use the virtual test bed and to control the devices freely. No further assistance was provided in order to allow for intuitive user utterances. We terminated a test run after approximately 22 commands on average depending on the duration of a complete session (15 min). In total, 900 spoken commands were recorded for all groups. The most important objective metrics we investigated are the number of understandings, non-understandings, and misunderstandings. Given the nature of the test system being a pure command-and-control SDS, metrics such as task-completion and dialogue success are less relevant in this context. In the following we provide detailed information on the dialogues we developed.

Table 5.5 The confirmation dialogue, the system generates automatically

Speaker	Utterance	Reaction
Suki	Please, the TV, could you switch it on?	–
Julia	–	[Keyword1 = TV] [Keyword2 = on]
Julia	Do you want to switch the TV on?	–
Suki	Yes!	–

5.6.1.1 Test Dialogues

The evaluation did not consist of a comprehensive dialogue as deployed during previous evaluations. Instead, the subjects were asked to utter commands they assumed to be suited to control the devices within the virtual test bed (see Fig. 5.21). In order to avoid misunderstandings, we implemented a confirmation dialogue that is initiated whenever one of the three approaches detects a spoken command that is not covered by the grammar before. An example of such an automatically generated dialogue is presented in Table 5.5. All sessions have been entirely protocolled and all user commands were separately recorded using a high-quality microphone.

During the implementation phase of the three understanding approaches we revealed a technical issue. A VoiceXML-based SDS provides low-quality 8 kHz wav files, which is appropriate for telephony applications. Contrary to this, the Microsoft Speech API requires at least 44.1 kHz of audio data so as to reach its maximum accuracy especially for OOV recognition tasks. Several options are conceivable to cope with this problem: the dialogue description standard (in our case VoiceXML) may be extended to support high-quality encoding of audio data or a German OOV recogniser with a 8 kHz corpus can be utilised. A further option would be to separately record the user input and pass it to the recogniser. We assumed this solution to be the most practical one. Thus, we decided to conduct the sessions with the external recogniser running in parallel. In the following section we take a look at the details of the questionnaire before we present the results of our investigation in Sect. 5.6.2.

5.6.1.2 Questionnaire

In this evaluation we focus on three objective measures: the number of understandings, non-understandings, and misunderstandings. Furthermore, the subjects were asked to fill in a questionnaire. We asked for personal details so as to draw a picture of the subjects and their (technical) background. Furthermore, the subjective user estimation of the system may reveal tendencies in terms of likes and dislikes regarding SDSs. Besides the questions about the personal details that are presented in Sect. 5.6.1, the subjects were asked to fill in a tailored SASSI questionnaire (Hone and Graham 2000) on a scale from zero (strongly disagree) to seven (strongly agree). We investigated the following measures:

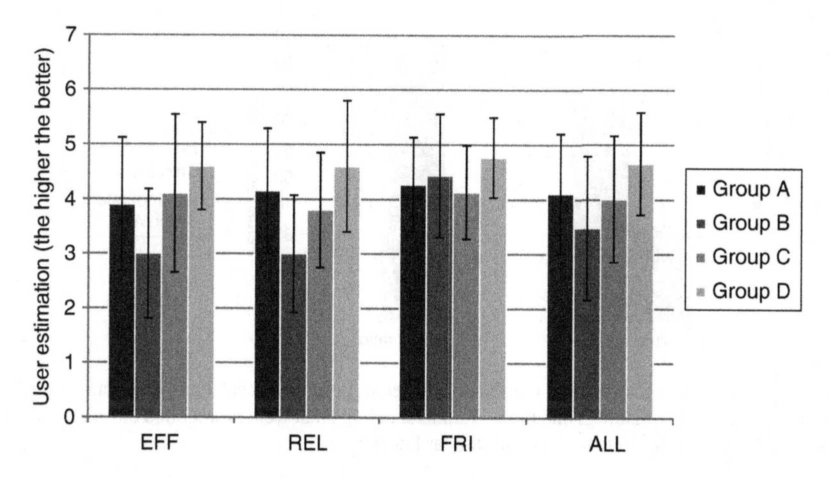

Fig. 5.26 The subjective results of the questionnaire on average together with the standard deviation

EFF *Efficiency* shows how efficient the subjects estimate the system to be, i.e. how well the dialogue can be comprehended by the user. This primarily relates to the confirmation dialogues presented in the previous section.

FRI *Friendliness* describes how user-friendly the system is and how pleasant it is for the subject to interact with the system.

REL *Reliability* shows how reliable the system is regarding mistakes and understanding issues.

ALL *Overall* reflects an overall subjective estimation by the user of the entire SDS.

Due to the limited number of test persons, the subjective results are meant to be a first indication, i.e. course estimation. In the following section we analyse the subjective and the objective results of the evaluation and investigate how the results influence our research.

5.6.2 Evaluation Results

Figure 5.26 plots the subjective user estimations for all groups on average. The approaches are measured neutrally (4) with slight deviations. The average values show a tendency that Group B, who used the *keyword-based* approach, rated the system worse than the other groups. Concerning EFF (efficiency) and REL (reliability) these differences are significant.

The Kruskal–Wallis test calculates a P value below 0.05 for both measures. The results indicate that concerning EFF and REL the *semantic-keyword* approach outperforms the other methods. The subjective estimation of user-friendliness did

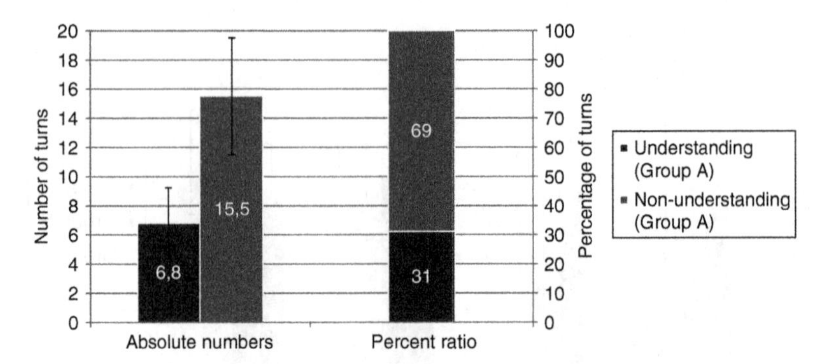

Fig. 5.27 The absolute average numbers (clustered columns) and the percentage (stacked columns) of the understandings and the non-understandings that were protocolled during the Group A session. The error indicators show the standard deviation

not show significant differences. The combined subjective measure (ALL) gives an indication of the overall system estimation. This group of bars summarises the subjective rating for all users: The *keyword-based* approach was rated worst, the *baseline* and the *blurred-keyword* approach were rated neutrally, and the *semantic-keyword* approach was rated best. The Kruskal–Wallis test calculates a P value of 0.04 indicating that the differences are statistically significant.

During an initial test phase, ten subjects used the system without any enhancements (Group A). In the following this group is referred to *baseline*. In total, the Group A users uttered 223 commands. Only 68 commands (31%) were recognised by the system. This weak result was expected and conditioned by the evaluation set-up: the users were asked to *intuitively* use the spoken command system. Since the three approaches extended the grammar-based system, we were able to prove the initial results during all evaluation sessions: the built-in grammar covered 33% of the spoken input during the *keyword-based* session, 26% during the *blurred-keyword* test run, and 30% during the evaluation of the *semantic-keyword* approach. These numbers underpin that a simplistic spoken command-and-control system, supporting limited functionality, cannot be used easily and intuitively. The objective results of the social evaluation point into a similar direction. The results indicate that even if the subjects receive a short introduction and are aware of the possible commands, it is a challenge for them to interact with an IE via voice. Figure 5.27 depicts the absolute average numbers of turns that were correctly understood. The system was able to recognise 6.8 turns on average. However, 15.5 turns were not understood by the ASDM. All turns have been protocolled and analysed after each session. Hence, we were able to preclude any unrelated input from our investigation. The system correctly rejected all unrelated input. The results clearly show that the baseline system (without any enhancements) and the dialogue models that provided minimal grammars do not adequately render the voice interface for intuitive usage (see Table 5.4).

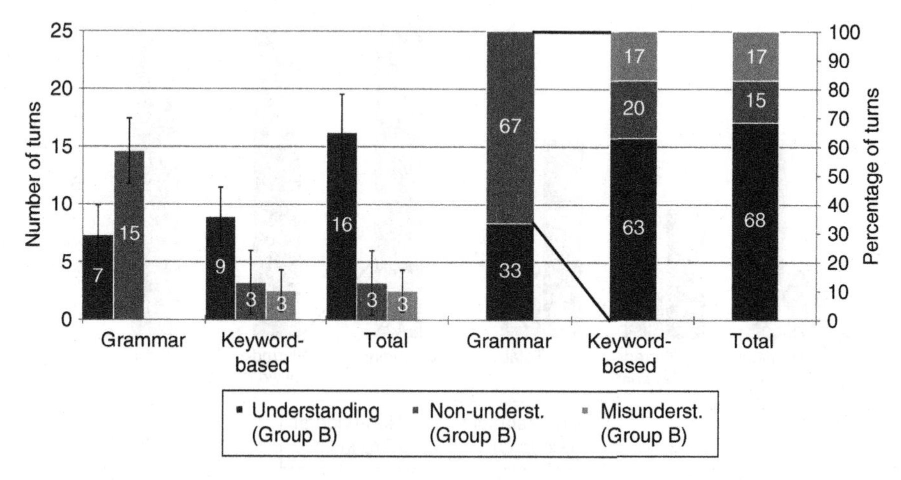

Fig. 5.28 The absolute average numbers (clustered columns) and the percentage (stacked columns) of the understandings, non-understandings, and misunderstandings that were protocolled during the Group B session. The error indicators show the standard deviation

Figure 5.28 plots the objective results that have been collected during the Group B session—the evaluation of the *keyword-based* approach. In total, 219 utterances were recorded during this session. Seventy-three (33%) utterances were covered by the grammar. On average, the prescribed grammars correctly matched seven utterances per user. These results are similar to the results of the *baseline* session. During the *keyword-based* test units, the ASDM was able to recognise 63% of the utterances that were not matched by the grammar before. On average, nine additional utterances were correctly understood. However, we still received 20% of non-understandings. We also recorded three misunderstandings per user on average (17% of all user utterances that were not covered by the grammar). In Fig. 5.28, the connectors between the stacked columns demonstrate the recognition process. The stacked column presenting the *keyword-based* results illustrates the recognition of user utterances that have not been recognised before. In total, the perceptual average rate of correctly understood user commands during this session was 68%. Overall we recorded 149 understandings, 37 misunderstandings, and 33 non-understandings during the session. Compared to the *baseline* approach the improvement is significant. The Mann–Whitney U-test calculates a P value of 0.008 disproving the hypothesis that the two results are statistically equal.

The second enhancement we analysed is the *blurred-keyword* approach that utilises a comparison under a specific Levenshtein distance. User utterances within this distance may match a specific keyword. Figure 5.29 shows the results of the Group C session. Two hundred and twenty-one commands were recorded during this evaluation session. Fifty-seven utterances (26%) were covered by the prescribed grammar. While the external recogniser as part of the keyword-based approach utilised the keywords as grammars, the blurred-keyword set-up made use of an OOV recogniser. Sixty-nine per cent of the previously not recognised

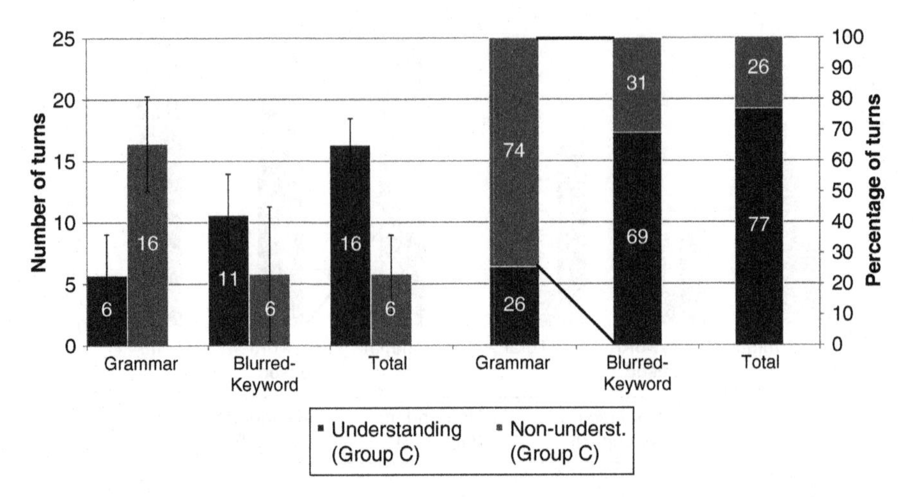

Fig. 5.29 The absolute average numbers (clustered columns) and the percentage (stacked columns) of the understandings and non-understandings that were protocolled during the Group C session. The error indicators show the standard deviation

utterances were correctly understood using the high quality audio data (indicated by the line connectors between the staked columns in Fig. 5.29). On average, 16 user commands were recognised by the ASDM. Still six non-understandings remained on average, eventuating in a total understanding rate of 77% (163 correctly understood commands). Compared to the *baseline* approach the *blurred-keywords* improved the understanding rate by 46%. The Mann–Whitney U-test calculates a P value of 0.004 underpinning the significance of the improvement between the grammar-based recognition and the *blurred-keyword* approach. Especially within a scaled up scenario, the *blurred-keyword* approach may be beneficial compared to the keyword-based set-up. An advantage is the higher coverage of possible user utterances by the fuzziness of the keyword comparator. Furthermore, a proper (and cognitive demanding) selection of keywords is less important since the Levenshtein comparison *corrects* understanding errors up to a specific limit. A too fuzzy comparison would increase the rate of misunderstandings. We therefore applied a Levenshtein distance for the comparison of possible keywords and user utterances of $\leq$ one. In the following we refrain from such a comparison but apply a semantic-lexical analysis of the user input by utilising a detection of semantic relations between the knowledge of the ASDM and the user input.

Figure 5.30 depicts the results of the *semantic-keyword* approach. Here, the user input was analysed by detecting semantic similarities between the domain keywords and the user utterance. During this test run we collected 237 commands. The grammar correctly matched 72 utterances. On average, the user input was correctly recognised seven times, while 17 commands could not be mapped appropriately. As with the blurred-keyword approach, we recorded no misunderstandings during this session. The semantic-keyword approach obtained a significant improvement

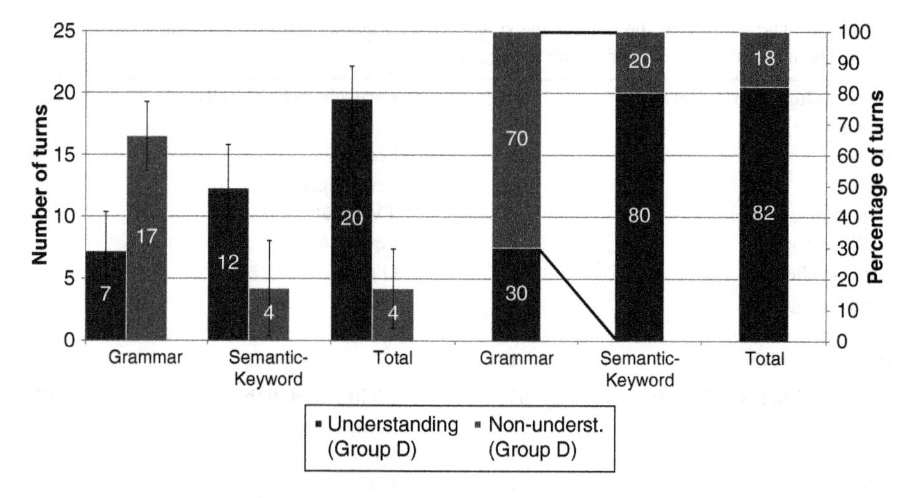

Fig. 5.30 The absolute average numbers (clustered columns) and the percentage (stacked columns) of the understandings and non-understandings that were protocolled during the Group D session. The 8 kHz data have been analysed online during the session and the 44.1 data offline, after the session has been finished. The error indicators show the standard deviation

resulting in 12 commands out of the 17 utterances that were mapped to the correct domain keyword. In total, 20 understandings and four non-understandings were recorded on average during the Group D session. Finally, the simplistic grammar, the OOV recognition combined with the semantic-lexical database GermaNet achieved a recognition rate of 82%. Compared to the *baseline* approach, the *semantic-keywords* improved the understanding rate by 51%. As with the previous session, the Mann–Whitney U-test shows a low P value of 0.002 underpinning the significance of the improvement between the grammar-based recognition and the *semantic-keyword* approach.

In absolute numbers, the combination of the semantic-lexical knowledgebase and the prescribed keywords led to additional 123 utterances that were correctly recognised. We estimate that the usage of the OOV recogniser (with high quality audio data) combined with the dynamic extension of keywords is beneficial especially within a larger domain yielding more commands and a more complicated information exchange between user and system. The better subjective user rating of the *semantic-keyword* approach concerning the efficiency and the reliability indicates that the subjects felt more comfortable with this approach. A reason could be that GermaNet allowed a broader usage of terms that, in practice, led to a more natural behaviour. This benefit is important with respect to colloquial utterances, e.g. "Glotze" (Eng. "box") for television set.

Such colloquial naming of devices that should be controlled via voice is rather rare. However, an SDS that understands this naming is perceived to be more natural and intuitive. Table 5.6 shows the naming of the six devices the subjects used in German language. The most common identifiers (marked as bold with their English translation) are most frequently used. Several uncommon names are also rarely

Table 5.6 The naming of the devices the subjects used together with the frequency they uttered them. The most common identifiers are marked as bold

Stereoanlage (Engl. stereo)	**64**	**Fernseher (Engl. telly)**	**141**
Musik	51	TV	6
Radio	41	Fernseh	6
Anlage	11	Fernsehgerät	4
Musikanlage	4	TV-Gerät	3
Radioreceiver	2	Fernsehen	1
Lautsprecher	1	Glotze	1
Audioanlage	1		
CD-Player	1		
Radiolautsprecher	1		
Jalousie (Engl. sun-blind)	**62**	**Licht (Engl. lights)**	**91**
Rolladen	58	Lampe	56
Rollo	25	Stehlampe	12
Vorhang	1	Leuchte	2
Vorhang	1	Stehleuchte	1
Ventilator (Engl. fan)	**88**	**Heizung (Engl. heating)**	**139**
Lüfter	9	Heizkörper	8
Gebläse	2	Radiator	3
Lüftung	1	Wärmequelle	1

used (e.g. "Audioanlage" for the English word "stereo"). It is not trivial to develop a grammar that covers such a variety of names for a higher number of devices and services, especially within a multi-tasking set-up. A further problem of large grammars is their maintenance. A main criterion how to choose the commands, a large-scale grammar consists of, is that the utterances should not sound similar so as to avoid misunderstandings.

Lightweight grammars can fulfil this important requirement because of their lower complexity. However, for larger and especially dynamically growing grammars it may be an issue to fulfil that requirement. This evaluation shows that a staged approach utilising a grammar-based fundamental understanding capability and a more intelligent extended recognition mode is beneficial. The former is meant to be used by experts (i.e. users that use the system regularly). The latter may intervene to understand intuitively uttered input. In the following section we summarise the approach, compare the three presented approaches, and draw some conclusions.

5.6.3 Conclusion

In this section, we introduced an extensive evaluation of three methods that can be used to improve the understanding capabilities of an SDS and, more concretely, of the ASDM. We divided the test field into four groups each containing ten subjects. The main task of all subjects was to *intuitively* control a virtual IE by the use of

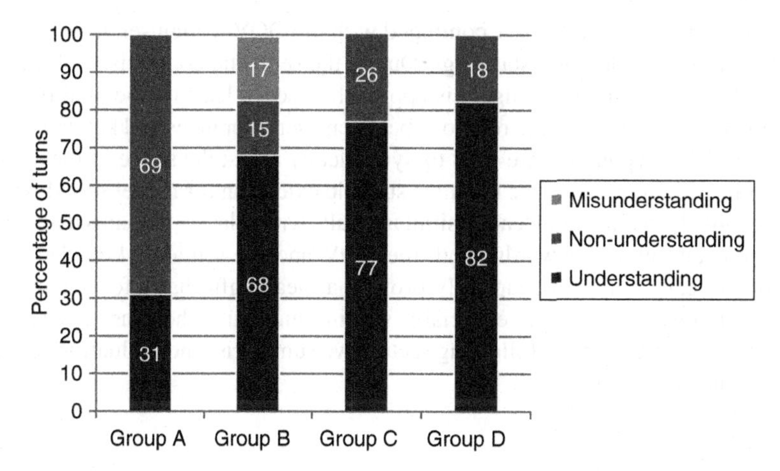

Fig. 5.31 The percentage of the understandings and non-understandings that were protocolled during the evaluation per group

spoken commands. As expected, the lightweight grammars only matched a few utterances. Figure 5.31 shows a recognition rate of 31% during the Group A session on average. Group A used a similar set-up as the other groups did, but no recognition enhancement was activated.

Figure 5.31 shows the Group B result. The system correctly interpreted 68% of the user commands. Group C and D performed better. The *blurred-keyword* approach achieved 77% and the *semantic-keyword* method 82% of correctly recognised utterances. Such a significant improvement strongly depends on the quality of the prescribed keywords. The *keyword-based* method mostly depends on the quality of the keywords. For a scaled-up version of the system, we assume that the task of selecting appropriate keywords is not trivial. During the evaluation, Group A received several misunderstandings. Seventeen per cent of the utterances caused a false-positive. This relates to the bad subjective average rating of the *keyword-based* approach regarding efficiency and reliability.

The presented approaches avoid the necessity of commands to be repeated by the user (in case of a non-understanding). Furthermore, they avoid misunderstandings by generating confirmative questions and by dynamically changing the applied understanding method. By using keywords and by segmenting the user input, it is possible to understand the user intention without comprehending the entire utterance. We have integrated several keywords into the system for detecting the domains and for finding the corresponding commands. These keywords may be utilised by the system to analyse a user utterance in case the built-in lightweight grammars fail in matching the spoken input correctly.

Due to the fuzziness of the *blurred-keywords*, they cover more variations of the prescribed keywords, thus making the approach more flexible. The *semantic-keyword* approach, however, outperforms the blurred-keywords. The subjects rated this method positively compared to the other approaches. By utilising the GermaNet semantic-lexical knowledgebase combined with the OOV recogniser, it receives the lowest number of non-understandings. Due to the semantic comparison of the user input with the system knowledge this approach depends least on the quality of the predefined keywords. Specific relations between user utterances and keywords (e.g. hypernyms and hyponyms) are used to dynamically adjust the scope of the system. We assume this approach to be the most suitable extension for model-based ASDM since it is likely that it would lead to similar results within larger domains using more commands and more keywords. Both the OOV and the semantic knowledgebase provide the capabilities to adaptively provide a meaningful basis for enhancing a classic, grammar-based ASR especially within situations where users *intuitively* interact with an SDS. In the following section we summarise the evaluation sessions and conclude this chapter.

5.7 Summary of the Evaluation Activities

In this section we summarise the outcome of the evaluation series that have been conducted with the OwlSpeak ASDM. We established the experiments so as to cover the most important levels of dialogue adaptation that are discussed in Sect. 3.6: Behavioural Adaptation, Dialogue Strategy Adaptation, Device Adaptation, Event Adaptation, and Task Adaptation. In the following we present an overview on the results and their relation to the specific adaptation level.

Initial evaluation In Sect. 5.1 we reported on the initial system evaluation. The main ASDM functionalities related to this evaluation are multitasking and dialogue control. These functionalities are defined as part of Task Adaptation and Behavioural Adaptation. The main outcome of the initial system is that the OwlSpeak ASDM runs stable and can be deployed within a real-life context. We successfully tested the (technical) combination of two independent but interrelated dialogue tasks. One of the most important results was that the applied multitasking capabilities lead to a cognitive overload on part of the user: strict multitasking without any system assistance proved to be inefficient.

Scalability analysis In Sect. 5.2 we discussed the results of a scalability analysis of the OwlSpeak ASDM. The fundamental functionality that the ASDM must be scalable relates to integratability. Thus, it is defined as part of Event, Device, and Task Adaptation. The experiment shows that the system in general can be scaled up to 100 SDOs for different devices and services. However, a computational issue is the ambiguity check that is conducted to analyse the user input. This pair-wise comparison of grammars leads to a computational time within $\mathcal{O}(n^2)$. Here, alternative approaches must be investigated in order to avoid unnecessary comparisons.

Topic switching strategies In Sect. 5.3 we reported on the different strategies to assist topic switching during an ongoing dialogue. The main ASDM functionalities related to this evaluation are multitasking and topic switching. These are defined as part of Event Adaptation, Device Adaptation, and Dialogue Strategy Adaptation. Motivated by the results of the initial system evaluation, we investigated enhanced methods of dialogue switching. A main result of this experiment is that specific strategies significantly improve the dialogue performance and allow for a lower cognitive load on part of the user. In total, the *explanation* strategy turns out to be the most promising approach. It was estimated by the users to be reliable and to provide a clear separation to an ongoing task. Furthermore, it has a positive impact on the dialogue quality in general.

Repair strategies In Sect. 5.4 we investigated possible repair strategies that can be applied to the ASDM. The functionality to apply repair strategies to an ongoing dialogue is defined as part of Dialogue Strategy Adaptation. A main result of the conducted investigation is that the selection of the "best" strategy is relative to the characteristics of the user, for example, to the experience with computer interfaces. While experts prefer intelligent repair strategies such as the *modified MoveOn* that is presented in Sect. 3.6.2.2, novices are irritated by such sophisticated approaches.

Qualitative social evaluation In Sect. 5.5 we reported on the social evaluation that has been conducted in the iSpace IE. The main functionalities concerning this qualitative evaluation are integratability and multitasking. These functionalities are defined as part of Device Adaptation. A main result of the conducted evaluation is that OwlSpeak proved to be integrable into an existing IE. The qualitative results indicated that experienced users were able to utilise the ASDM successfully—and enjoyed using it. However, for novices the system revealed to be awkward. The results showed a high percentage of user utterances the system did not interpret correctly. The lightweight grammars did not allow for an *intuitive* usage of the system.

Advanced understanding methods In Sect. 5.6 we focussed on the advanced understanding methods that were applied to the system so as to encounter the issues regarding intuitiveness we were confronted with during the social evaluation in the iSpace. The main ASDM functionalities related to this evaluation are adaptive understanding and dialogue control. These functionalities are defined as part of Speech Adaptation and Behavioural Adaptation. During the evaluation we compared three understandings enhancements using an external recogniser with the baseline set-up of the OwlSpeak ASDM. The results indicate that an additional keyword-based recognition significantly enhances the system's recognition capabilities. The enhancement strongly depends on the quality of the keywords. For this reason, the *semantic-keyword* method that achieved the best result would be the most appropriate enhancement. Here, the system behaviour does not solely depend on the quality of the keywords and even colloquialisms may be recognised correctly.

In summary, we proved that the OwlSpeak ASDM provides a fully functional framework for adaptive spoken dialogue within the context of IEs. The evaluation series show that many issues can be successfully faced by extending the underlying knowledgebase—the SDOs. The strict distinction of the dialogue logic, the knowledgebase, and the generated dialogues comes to a very flexible and at the same time powerful basis that allows for dialogue adaptation regarding several levels. These levels of adaptation were defined as part of this work and describe the most important functionalities required to cope with the three main stakeholders that interact with the ASDM: the environment (i.e. the IE), the SDS (that may be error-prone), and user himself.

Chapter 6
Conclusion and Future Directions

We have developed and evaluated a formal framework for ASDM. The OwlSpeak system fulfils several requirements that arise from (multi-)task-based situations that especially occur within IEs. We have investigated seven levels of adaptation for three stakeholders that employ spoken dialogue. These are: the user, the SDS, and the IE. By integrating the ASDM in the centre of these parties, it acts as a mediator establishing the interface. Different levels of adaptation arise for each of the stakeholders.

We derived two levels of adaptation that can be applied to ASDM in order to facilitate a smooth interaction with the user: Emotional Adaptation and Behavioural Adaptation. The former may be realised by allowing for speaker aware dialogues. Such dialogues react to the user's mood or other personal characteristics. We described details about this adaptation type in Sect. 3.1 and the results have been published in Schmitt et al. (2009). The latter level of adaptation describes the possibilities that the user has to directly influence the dialogue through "special behaviour". We defined this special behaviour to be dialogue related. Examples for this level of adaptation are signalling-type phrases such as "don't bother me", signalising that the user does not want to speak to the system at the moment. Thus the ASDM must support specific functionality to provide the user with direct control over the dialogue.

The inherent properties of the architecture and the model-based approach we have chosen provide direct user influence over the ongoing dialogue and on the dialogue models in general (see Sects. 3.6.5 and 4.2.1). The second stakeholder—the SDS itself—implies two levels of adaptation to be realised by the ASDM: Dialogue Strategy Adaptation and Speech Adaptation. Since one of the most important components of an SDS, the ASR, is error-prone, it is mandatory to provide the functionality to adapt the dialogue strategy so as to avoid harmful dialogue situations or to repair the dialogue (usually) with the help of the user. We have shown that the ASDM is able to incorporate different dialogue strategies (see Sects. 3.6.2, 3.6.2.1, and 3.6.2.2) and presented the outcome of an in-depth comparison of different topic switching strategies in Sect. 5.3. The results have also been published in Heinroth et al. (2011).

T. Heinroth and W. Minker, *Introducing Spoken Dialogue Systems into Intelligent Environments*, DOI 10.1007/978-1-4614-5383-3_6,

Furthermore, different repair strategies have been evaluated in Sect. 5.4. The second level of adaptation that has been defined to be related to the SDS is Speech Adaptation. We were able to significantly enhance the understanding capabilities of OwlSpeak by integrating a second recogniser that provided OOV recognition if it failed on its first attempt. We have compared different methods of detecting user utterances and we found coherences between the utterances and the contents of the dialogue models (see Sects. 3.6.3 and 5.6). As a part of this document we have also defined adaptation regarding a third stakeholder: the IE that integrates the ASDM. We defined three levels of adaptation to describe the requirements of the ASDM and thus its desired capabilities. Here we distinguish between Event, Device, and Task Adaption (Heinroth et al. 2010).

In Sect. 3.3 we defined these three levels. The ASDM is able to pause and resume active tasks, add and remove dialogue domains, permanently save the state of a dialogue, and to provide more than one active spoken dialogue in parallel. Hence, many of the requirements of the IE have already been fulfilled. The main functionality—support of multitasking for spoken dialogue—is of major importance for responding to urgent events that may occur at any time and notably during an ongoing dialogue. In order to integrate the control of several devices and services into *a single* voice interface, it is mandatory for coping with a variety of tasks. Therefore, the functionality to generate combined spoken dialogues from different dialogue models (SDOs—see Sect. 4.3) is one of the outstanding characteristics of the OwlSpeak ASDM. The definition of the SDO and its framework will be discussed in the following section. Further details can be found in Heinroth and Denich (2011).

The different aspects of adaptation did not only guide the implementation of our prototype design but also influenced our evaluation strategy. We investigated the overall system functionality and the interoperability of the dialogue models in Sect. 5.1. This initial system test revealed the issues users may have when the ASDM initiates a topic switch due to external events. The results motivated us to conduct an in-depth user evaluation regarding different topic switching strategies (see Sect. 5.3). The results revealed the so-called *explanation* strategy to be the most efficient way to perform topic switches in an ongoing dialogue. Our investigation will be suitable for future voice interface design (Heinroth et al. 2011).

Using the iSpace IE (University of Essex), we also investigated how users deploy the ASDM as part of an ATRACO system under realistic conditions during a social evaluation. The qualitative results indicated a serious general deficiency of SDSs: the interface is meant to be used in a natural way. This often leads to a specific degree of uncertainty in the system response. At the beginning of the tests most users request a system that responds more accurately. At the same time, the users went through a learning process over three sessions. At the end they were still motivated and also experienced a much better dialogue performance compared to the first session. This indicates that users are determined to use spoken dialogues in order to interface with an IE, even if they do not work perfectly.

During the implementation phase we incorporated certain data, such as the names of dialogue domains and lists of keywords, into the dialogue model. In order to respond to the results of the social evaluation we implemented several methods to enhance the understanding of OwlSpeak (see Sect. 4.2.3.4). Our study revealed that a semantic comparison was most suitable for finding user utterances that are not correctly recognised but that still relate to specific commands that the system is able to understand (see Sect. 5.6). For this purpose we integrated the GermaNet lexical-semantic database into the OwlSpeak ASDM (Heinroth et al. 2012).

We also investigated ways to cope with incorrect or even misunderstood user input: repair strategies (Zgorzelski et al. 2010). These strategies applied to the ASDM help to recover from an incorrect dialogue state. In this instance, the evaluation revealed which repair strategy performed better depends on the type of user (e.g. novice or expert SDS user). Finally, we conducted a system scalability analysis to determine how the ASDM can be used if up to 100 devices have to be controlled via voice. The results demonstrated that the system itself is scalable; however, a specific component—the *conflict resolution*—must be redefined if the number of devices and services is high. In this case, cross-comparison is the main difficulty. In the remainder of this chapter, we discuss our most important findings before describing directions for further research in Sect. 6.3.

6.1 Adaptive Spoken Dialogue Management Approach

Our work contributes to ongoing progress in the scientific field of Spoken Dialogue Management. In Sect. 2.4 we introduced several scientific approaches towards realising this crucial component that undoubtedly must be part of an SDS. However, most of the state-of-the-art approaches to SDM use complex rule bases or statistical models to define the dialogue logic. Both directions are promising as long as the dialogue content does not change dynamically during runtime (i.e. during an ongoing dialogue). This is a major drawback if the dialogue logic must cope with changing domains. We face the following challenge, especially within an IE: various domains, tasks, devices, and services populate such an environment. Therefore, the dialogue logic must be able to incorporate new dialogue descriptions into its knowledgebase and to decouple older ones.

One of our most important contributions is the description of dialogues by the use of Spoken Dialogue Ontologies. We defined this format as part of our work and utilised it for the definition of various different dialogues ranging from command-and-control forms to complex conversations for information exchange. An SDO is a set of specific OWL classes that formalise all information needed to describe a specific spoken dialogue (see Appendix B). The SDO has also been published as part of the Stanford Protégé Ontology Library.[1] As depicted in Fig. 6.1 the most important feature is a formal description of both dialogue state *and* structure.

[1]Visit http://protegewiki.stanford.edu/wiki/Protege_Ontology_Library and look for "Spoken Dialogue Ontology"

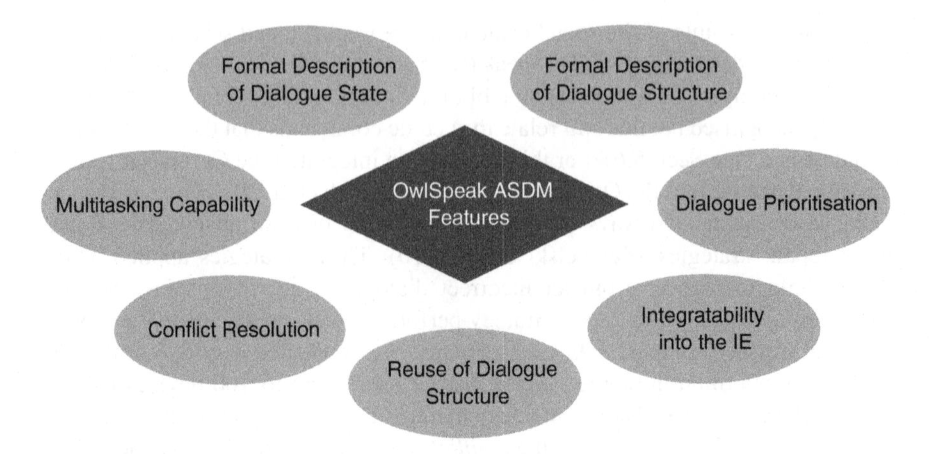

Fig. 6.1 The most important features of the OwlSpeak ASDM

This means that dialogue domain does not only consist of static information encoding the dialogue discourse but also evolves during the ongoing dialogue: it is able to store the dynamic information that describes the dialogue progression persistently. This unique feature allows for pausing dialogues and therefore to switch between different domains and tasks. So as to (re-)start a specific dialogue, the formal framework also offers the possibility of deleting all dynamic data while the static information is kept for the initialisation of the dialogue. One of the reasons to choose OWL as SDO format was that it is possible to import ontologies into related ontologies and furthermore to inherit class definitions together with their potential instances (i.e. objects) from specific "mother-ontologies".

This allows for a comprehensive reuse of already defined dialogue turns (see Sect. 3.5.2) or of complete dialogue constructs. The dialogue logic that works on these formal definitions is responsible for the decision of which dialogue turns have to be used for the generation of the conversation. The current prototype uses a set of prescribed rules and a prioritisation mechanism for judging this decision. This prioritisation can either be statically defined (e.g. by the dialogue designer) before the dialogue starts or can explicitly be influenced by spoken commands (e.g. by the user) during an ongoing dialogue.

A third option has been implemented that allows a dynamic change of the priorities of specific dialogue turns depending on the progress done in user–system interaction. An example for this behaviour would be a spoken reminder that becomes more urgent while the user talks to the system (e.g. the washing machine is done). In this case the system is able to automatically raise the priority of the reminder until it exceeds the main dialogue. Herewith, the main benefit is that we keep a chance the ongoing dialogue does not have to be interrupted while the system guarantees that the user would not miss the urgent event. In Sect. 6.3 we present several other approaches to intelligent dialogue prioritisation as part of the future directions.

Besides the prioritisation mechanisms, the multitasking capability of the Owl-Speak ASDM requires another important feature: the semi-automatic conflict resolution. Especially when it comes to various devices that have to be controlled within an IE, we have to face the issue that some devices provide similar commands. For example, two different lighting devices might usually understand similar commands such as "light on". It is then up to the ASDM to decide which light the user wants to switch on. For this issue, external information such as gestures, the current location of the user, and the viewing direction could all be taken into account as we outline these examples in Sect. 6.3.

However, for the prototype, a voice-based conflict resolution is provided by an automatically generated dialogue. This dialogue provides the informative names of the affected devices and asks the user to choose which device he wants to trigger. This method of conflict resolution seems to be one of the most promising options for solving this issue since it works semi-automatically. A major requirement for the integration of various devices and tasks is the interoperability of the ASDM. For this purpose we have defined a UPnP wrapper providing access to the most important features of the ASDM such as the integration of new dialogue descriptions (i.e. new SDOs), the reset of dialogues that must be re-initialised, and a "pause" command halting all dialogue, to name but a few. All these commands can be expressed via speech on the part of the user.

The complete dialogue management framework (OwlSpeak) is freely available from Sourceforge in order to allow a community-wide spread of the software for further research (cf. Sect. 4.5). We have also published several spoken dialogues described as SDOs as an example for dialogue designers. The theoretical contribution of a working ontology-based spoken dialogue description framework and the prototype implementation of ASDM provides a fertile ground for the evaluation sessions we conducted. In the following section we discuss the most important results.

6.2 Evaluation Results

The evaluation strategy followed two directions: the first target of our investigation was motivated by the question of whether the defined ASDM framework works correctly, stably, and reliably. The initial system test showed that our prototype fulfilled these fundamental requirements. The scalability of the system also played a crucial role during the technical parts of the evaluation sessions. For the initial system test we utilised SDOs with more than 200 individuals to describe complex dialogues for information gathering. As depicted in Fig. 6.2 we also investigated how the system performs if there are numerous devices and services involved that can be controlled via speech. An important outcome of this analysis is the scalability of the OwlSpeak ASDM. However, semi-automatic conflict resolution is computationally demanding when there are various options of what the user

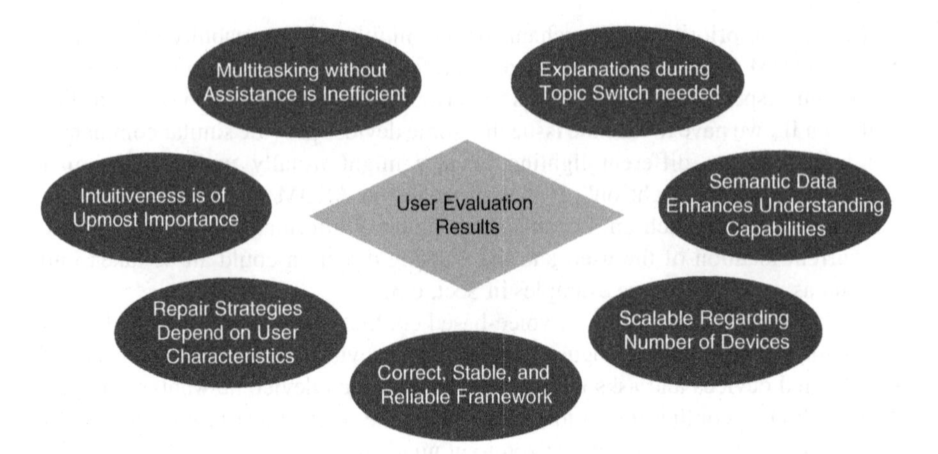

Fig. 6.2 The most important results of the OwlSpeak ASDM evaluation sessions

might utter (i.e. grammars) involved in the generation of the dialogue: the current prototype cross-compares these descriptions so as to detect potential conflicts. Since we assume that in practice the user would not interact with more than 30 activated devices and services at one particular time, we do not believe that the scalability analysis indicates a limitation of the framework regarding this issue.

The second main direction of our evaluation efforts were looking at specific characteristics of the ASDM that may have to be optimised. These were chosen to be generic in the sense that they can be transferred to any other SDMs or comparable components that provide the dialogue logic. Figure 6.2 lists the major blocks of the evaluation as follows:

- Multitasking
- Topic switching
- Intuitiveness
- Intelligent understanding
- Repair strategies

The multitasking capabilities of OwlSpeak were tested during the initial system evaluation. Multitasking is a broader concept. It is therefore necessary to split it into more concrete system functionalities. Hence a focus was set on the dynamic prioritisation strategy that leads to topic switches during the ongoing dialogue. While the concept was technically convincing, subjects did not want to be confronted with sudden switches of the topic as long as they were involved in a specific dialogue. This result led to a second evaluation series focussing on different strategies that can be applied in order to assist the user if the ASDM must switch the dialogue focus to a different topic (and afterward back to the original one). In total, 80 subjects tested different topic switching strategies and rated their performance. The analysis of the subjective results and objective measures such as the number of non-understandings and the dialogue duration indicated that a sophisticated strategy referred to as *explanation* outperformed the others. This strategy guided the users by informing them that the dialogue must be interrupted.

After the sub-dialogue is processed, the original dialogue is reintroduced by prompting the dialogue topic (e.g. "back to dinner preparation"). Notably, the applied dialogue strategy measurably and significantly influenced the overall dialogue quality. The social evaluation session that has been conducted within an existing IE under realistic conditions, on one hand, revealed some qualitative—and subjective—results that are of interest (see Sect. 5.5). On the other hand, it also revealed some objective results that exposed a main drawback of the approach. The subjective results relate to a positive learning process that the subjects went through during three evaluation sessions.

All subjects were motivated to use a voice interface within an IE. After the first session they were not disappointed: the objective results indicated a very high number of non-understandings. We also recorded a significant improvement during the second session and a further improvement during the third session. The ASDM was kept unchanged during all sessions. The integrated prototype provided straightforward command-and-control dialogue structures. However, these structures were obviously too rigid and insufficiently intuitive. Further results of the ATRACO evaluation and details on the prototype are published in Heinroth and Minker (2011). Taking the subjects' high motivation and eagerness to control the environment into account, we investigated ways to enhance the understanding capabilities of OwlSpeak. The main objective was to render the interface more intuitive.

Hence, during the fourth user evaluation we studied different mechanisms that can be applied to the OwlSpeak ASDM in order to enhance the understanding capabilities and at the same time keep the complexity of the domain models low. Notably, the subjects received no introduction during this evaluation on how to control a virtual IE via voice: they were forced to use the interface *intuitively*. We compared three mechanisms that potentially allow the system to understand more system-directed input than the strict grammars defined as part of the SDOs allow: a keyword-based approach, a fuzzy Levenshtein-based enhancement, and a semantic-keyword approach that utilises GermaNet (a semantic knowledgebase similar to WordNet).

Our investigation revealed the semantic-keyword approach to be most appropriate for significantly enhancing the understanding of a grammar-based SDM such as OwlSpeak. This approach addressed the user input that was semantically related to input described by the grammar defined as part of the SDOs. Here, our research points in the same direction as a commercial system that has recently been released for iOS and OS X Lion: Siri by Apple. This concept places the focus on semantic understanding rather than on exact recognition. We speculate that future SDSs will be measured by their *understanding capabilities* rather than by their *recognition rate*, which still is the most important measure for scientific and commercial systems.

The fifth evaluation session covered the question of how to cope with common mistakes that occur during a human–machine interaction via voice. Therefore, we compared three strategies ranging from a simple re-prompt to a more complex self-repair strategy that tried to guess the correct user input. This *guessing* was only possible if the dialogue consisted of interrelated information. The main outcome of this evaluation was that the repair strategy to be used strongly depends on user characteristics. The subjective rating of experts significantly differed from the rating of novices. This underpins the importance of user related information that might easily be integrated into the SDOs, i.e. the dialogue description OwlSpeak builds upon. The work presented in this document raises interesting questions and points to directions for future research. These will be discussed in the following.

6.3 Future Directions

Due to the modular and layered implementation of the OwlSpeak ASDM, various directions for improvements and future research are conceivable. In the following, for each of the main layers—model, view, and presenter—we outline possible directions for future work. The underlying idea of the *model* was to implement a format that can be utilised to store a spoken dialogue description and its corresponding state. We have enhanced this model by adding semantic information that is used to identify dialogue domains and, more specifically, to identify conversational acts and therefore enhance the natural language understanding capabilities of the ASDM.

Our starting point, however, was that the ASDM usually ran in a "grammar-based" mode: in that case semantic understanding enhancement is only activated if the system detects an understanding error. Regarding future work, semantic understanding could be set to "always activated" thus avoiding the use of grammars totally. We are confident that command-and-control dialogues would benefit from such an approach. More complex dialogue exchanges, such as negotiation or information retrieval on the part of the user, may still need special grammar to describe, for example, lists of city names or ingredients of recipes to be discussed with the user.

The model may also be further enhanced by adding information that is related to different modalities. Since the OWL format supports multiple languages, the OwlSpeak ASDM framework has been kept multilingual as well. Hence, all conversational acts can be defined in different languages. A similar strategy can be followed to define different constructs for different modalities on a turn-wise basis. Depending on which modality the user or the system chooses for interaction, the framework would be able to generate, for example, GUIs instead of VUIs. Therefore, it would be necessary to integrate user interface descriptions such as the Extensible Interface Markup Language (XIML, see Puerta and Eisenstein 2002) or the User interface eXtensible Markup Language (UsiXML, see Limbourg et al. 2005) into the SDO. Since the OWL format is open and highly flexible we assume that technically this would be straightforward. However, practical issues might arise when it comes to conflicts that may need to be solved in a way similar to the way that the OwlSpeak ASDM solves conflicts (cf. Sect. 4.2.3.2).

The second group of considerations concerning future directions relates to the second layer of OwlSpeak, to the *view*. For our prototype we have implemented two different types of views: a VoiceXML-based version providing spoken dialogue and an HTML-based format that can be used for testing during the phase of dialogue design. Building upon the future direction towards an SDO that provides information for generating multimodal dialogues, it would also be necessary to investigate the possibilities for generating multimodal views. An option would be to integrate the Extensible MultiModal Annotation markup language (EMMA, see Johnston et al. 2009) into the mechanisms of OwlSpeak that generate the view. Further research is needed regarding new types of dialogue turns, i.e. conversational acts, the system should be able to render.

As part of our work, we have introduced the *system turn*, the *user turn*, and the *exchange* to be the basic blocks that are used to formulate the dialogues (see Sect. 3.5.2). In principle, even more sophisticated dialogues consisting of so-called over-answering situations can be modelled using the framework (during an over-answering occurrence, the user provides information that relates to more than one conversational act). The current prototype, however, cannot generate such over-answering turns *automatically*. While the SDOs theoretically provide the data needed to grasp input on the part of the user that relates to different turns (i.e. "I want to fly *from* Frankfurt *to* Brussels.") the current view does not combine the *from* and the *to* turns automatically. We have already implemented a proof-of-concept regarding this issue but have not integrated the functionality into the stable version of OwlSpeak yet.

The third and obviously one of the most important directions of future work relates to the *presenter* layer, which is providing the dialogue logic. There are two major methods that may offer ways to enhance this layer. First, dialogue strategies that have been evaluated and manually added to the prototype must be integrated into the fundamental dialogue logic. We have integrated the detection of system initiated topic switches. This detection can be used to automatically generate (spoken) dialogues that provide proper topic switching strategies. These are applied *without* extra information that must be part of the SDOs.

Furthermore, it will be necessary to implement sophisticated functionalities that enable OwlSpeak to detect dialogue topic switches that are initialised by the user. The current prototype needs these switches to be implicitly triggered by the user (e.g. via voice). Future versions of OwlSpeak, however, may use semantic information extracted from the user input to automatically apply a dialogue topic switch to a different task. The SDOs also provide the information necessary to apply undo and redo functionalities. These techniques are also needed to integrate the tested repair strategies into the prototype.

Secondly, a different basis may provide the dialogue logic itself. The current prototype applies a mixture of rule- and priority-based mechanisms for generating spoken dialogues. While this approach leads to a stable (and computable) framework the underlying logic is quite rigid. Thus the current prototype only deals with the most appropriate user utterance as input. This means that only the result of the ASR with the highest probability is rated to be the correct input. The uncertainty of speech recognition has not been taken into account. We are currently working towards a set-up that integrates so-called Partially Observable Markov Decision Processes (POMDPs). A POMDP is a mathematical framework for planning and acting in an environment under uncertainty (Kaelbling et al. 1998).

In practice, POMDPs are often computationally intractable. Especially when it comes to a large number of states (i.e. dialogue variations) dialogue decisions cannot be provided within reasonable time periods. However, several approaches towards an approximation of POMDPs within the SDS context are presently available (Williams and Young 2007; Young et al. 2010; Habibi et al. 2010). By integrating the POMDP approach into OwlSpeak we may achieve two goals: (1) allowing the system to handle uncertain information and therefore to combine additional sources of information within the process of dialogue decision and (2) maintaining the computational tractability of the system by utilising the dialogue data provided by the SDOs.

In conclusion, a stable and reliable OwlSpeak ASDM will provide fertile ground for future R&D concerning the various levels of adaptation for the major stakeholders of adaptive spoken dialogue—the user, the SDS and, the Intelligent Environment itself.

Appendix A
Control Dialogues as Part of the ATRACO IE

A.1 Audio Player

As part of the Entertainment Sphere AS1 an audio player is integrated within the AS. This device can be controlled via a graphical interface and via the ASDM. After the IA makes this spoken interface available the user has two options to start the audio controlling dialogue: Either he may simply utter "music" or he activates the audio controlling widget on a screen at hand. After the spoken interface is activated the following commands can be used:

- "Play" to play the current song
- "Pause" to pause the current song
- "End" to terminate the music controlling dialogue
- "Next Song" to play the next song
- "A song back" to play the previous song
- "Crank up the volume" to turn the music's volume up
- "Turn the music down" to turn the music's volume down
- "Classic" to select the "Classic" playlist
- "Chill out" to select the "Chill out" playlist
- "Jazz" to select the "Jazz" playlist
- "Latin" to select the "Latin" playlist
- "Pop" to select the "Pop" playlist
- "Rock" to select the "Rock" playlist

The grammars to understand the various commands have been kept as simple as possible in order to delimitate complexity of the underlying dialogue. Several commands can be reused for controlling the photo frame presented in the next subsection. If the user (or the IA) has activated both spoken interfaces the ASDM automatically detects conflicts that might occur and initiates a sub-dialogue. It asks the user if the last command refers to music control or to the interface of the photo frame. After the user has declared his intention by answering, for example, "music" the command can be executed. Of course, the grammars listed above are only

T. Heinroth and W. Minker, *Introducing Spoken Dialogue Systems into Intelligent Environments*, DOI 10.1007/978-1-4614-5383-3,

examples that represent a minimal setup. If necessary all grammars can be extended as required. For example, if the user wants to avoid conflicts, he might utter “Next song” instead of “Next” and thus does not has to select the corresponding dialogue domain if the photo frame interface is activated as well.

A.2 Photo Frame

Similar to the audio player a digital photo frame also takes part of AS1. It is also accessible via the graphical interface or via the ASDM. Accordingly to the audio player interface the user may utter “photos” to initiate the dialogue in case the IA hasn’t already activated it. After the spoken interface is activated the following commands can be used:

- “Play” to activate the auto-slideshow
- “Pause” to pause the slideshow
- “End” to terminate the photo dialogue
- “Next” to show the next photo
- “back” to show the previous photo
- “France” to select the “France” slideshow
- “Italy” to select the “Italy” slideshow
- “Slovenia” to select the “Slovenia” slideshow
- “Switzerland” to select the “Switzerland” slideshow
- “Venice” to select the “Venice” slideshow
- “Wedding” to select the “Wedding” slideshow
- “Photo details” to request the photo details

The dialogue interfaces to the audio player and to the photo frame are quite similar. We aimed at investigating if users cope or even like the automatically conflict solving capability of the ASDM. If the user utters “Photo details” he receives the title of the picture that is currently shown on screen as answer.

A.3 Lights

All ATRACO Activity Spheres support light control. The Fuzzy Task Agent (FTA), as described in detail within Wagner and Hagras (2010), takes care of the actual level of light. Users are able to influence the decisions of the FTA by using the graphical interface and/or the ASDM. The following commands can be used for this purpose:

- “light very low”
- “light low”
- “light medium”

- "light high"
- "light very high"
- "light on"
- "light off"

We have decided to use a grammar based on linguistic labels (veryLow, low, medium, high, veryHigh) for controlling the lights instead of, for example, percentages. The FTA can integrate these values into his knowledgebase in order to adjust the light to the user's individual preferences. In other words, the meaning of a low lighting level differs from one user to another. Since the FTA learns what "low" means for a user the usage of exact percentages would override this kind of intelligent behaviour.

A.4 Heating and Alarm

Both heating and alarm control incite similar dialogues. The user may utter "heating on", "alarm on", "heating off", and "alarm off". From a spoken interaction point of view these dialogues are less interesting. However, these dialogues show that the proposed framework provides methods to easily develop and integrate new dialogues. The Spoken Dialogue Ontologies used to generate the alarm and the heating interface only consist of seven individuals.

A.5 Yes–No-Question

The Yes–No-Question is used to retrieve information from the user. More precisely, if one of the components of the ATRACO system needs a specific input from the user this dialogue may be initiated. In order to keep the complexity reasonable we have decided to only support queries that can be answered with "yes" or "no". If the user does not want to or cannot provide input he may also utter "leave it". Since the meaning and the content of the question can be set during run-time the yes–no-question dialogue serves as a template that has to be adapted by the specific component that needs input before the dialogue can be initiated by the IA and provided by the ASDM.

Appendix B
The Spoken Dialogue Ontology

In this chapter we present the Spoken Dialogue Ontology (SDO) that is used as model describing both state and structure of spoken dialogues. The OWL ontology has been published as part of the Protégé Ontology Library.[1]

```
1   <?xml version="1.0"?>
2   <rdf:RDF
3       xmlns:rdf="http://www.w3.org/1999/02/22-rdf-syntax-ns#"
4       xmlns:protege="http://protege.stanford.edu/plugins/owl/protege
          #"
5       xmlns:xsp="http://www.owl-ontologies.com/2005/08/07/xsp.owl#"
6       xmlns:owl="http://www.w3.org/2002/07/owl#"
7       xmlns="http://localhost:8080/Atraco/OwlSpeakOnto.owl#"
8       xmlns:xsd="http://www.w3.org/2001/XMLSchema#"
9       xmlns:swrl="http://www.w3.org/2003/11/swrl#"
10      xmlns:swrlb="http://www.w3.org/2003/11/swrlb#"
11      xmlns:rdfs="http://www.w3.org/2000/01/rdf-schema#"
12    xml:base="http://localhost:8080/Atraco/OwlSpeakOnto.owl">
13    <owl:Ontology rdf:about=""/>
14    <owl:Class rdf:ID="History">
15      <rdfs:subClassOf>
16        <owl:Class rdf:ID="State"/>
17      </rdfs:subClassOf>
18    </owl:Class>
19    <owl:Class rdf:ID="Utterance">
20      <rdfs:subClassOf>
21        <owl:Class rdf:ID="Speech"/>
22      </rdfs:subClassOf>
23    </owl:Class>
24    <owl:Class rdf:about="#State">
25      <rdfs:subClassOf>
26        <owl:Class rdf:ID="DialogueDomain"/>
27      </rdfs:subClassOf>
28    </owl:Class>
29    <owl:Class rdf:ID="Move">
30      <rdfs:subClassOf>
31        <owl:Class rdf:about="#Speech"/>
32      </rdfs:subClassOf>
33    </owl:Class>
34    <owl:Class>
35      <owl:unionOf rdf:parseType="Collection">
```

[1] http://protegewiki.stanford.edu/wiki/Protege_Ontology_Library

T. Heinroth and W. Minker, *Introducing Spoken Dialogue Systems into Intelligent Environments*, DOI 10.1007/978-1-4614-5383-3,

```
36        <owl:Class rdf:about="#Move"/>
37        <owl:Class rdf:ID="Agenda"/>
38      </owl:unionOf>
39    </owl:Class>
40    <owl:Class rdf:ID="BeliefSpace">
41      <rdfs:subClassOf>
42        <owl:Class rdf:ID="Belief"/>
43      </rdfs:subClassOf>
44    </owl:Class>
45    <owl:Class>
46      <owl:unionOf rdf:parseType="Collection">
47        <owl:Class rdf:about="#Belief"/>
48        <owl:Class rdf:about="#Move"/>
49      </owl:unionOf>
50    </owl:Class>
51    <owl:Class rdf:about="#Belief">
52      <rdfs:subClassOf rdf:resource="#State"/>
53    </owl:Class>
54    <owl:Class rdf:ID="Semantic">
55      <rdfs:subClassOf>
56        <owl:Class rdf:about="#Speech"/>
57      </rdfs:subClassOf>
58    </owl:Class>
59    <owl:Class rdf:about="#Agenda">
60      <rdfs:subClassOf rdf:resource="#State"/>
61    </owl:Class>
62    <owl:Class>
63      <owl:unionOf rdf:parseType="Collection">
64        <owl:Class rdf:about="#Belief"/>
65        <owl:Class rdf:about="#Move"/>
66      </owl:unionOf>
67    </owl:Class>
68    <owl:Class rdf:ID="Grammar">
69      <rdfs:subClassOf>
70        <owl:Class rdf:about="#Speech"/>
71      </rdfs:subClassOf>
72    </owl:Class>
73    <owl:Class rdf:ID="WorkSpace">
74      <rdfs:subClassOf rdf:resource="#Agenda"/>
75    </owl:Class>
76    <owl:Class rdf:about="#Speech">
77      <rdfs:subClassOf rdf:resource="#DialogueDomain"/>
78    </owl:Class>
79    <owl:Class rdf:ID="Variable">
80      <rdfs:subClassOf rdf:resource="#Speech"/>
81    </owl:Class>
82    <owl:ObjectProperty rdf:ID="has">
83      <rdfs:domain rdf:resource="#Agenda"/>
84      <rdfs:range rdf:resource="#Move"/>
85    </owl:ObjectProperty>
86    <owl:ObjectProperty rdf:ID="semantic">
87      <rdfs:domain>
88        <owl:Class>
89          <owl:unionOf rdf:parseType="Collection">
90            <owl:Class rdf:about="#Move"/>
91            <owl:Class rdf:about="#Belief"/>
92          </owl:unionOf>
93        </owl:Class>
94      </rdfs:domain>
95      <rdfs:range>
96        <owl:Class>
97          <owl:unionOf rdf:parseType="Collection">
98            <owl:Class rdf:about="#Semantic"/>
99            <owl:Class rdf:about="#Variable"/>
100         </owl:unionOf>
101       </owl:Class>
102     </rdfs:range>
```

```
103     </owl:ObjectProperty>
104     <owl:ObjectProperty rdf:ID="utterance">
105       <rdfs:range rdf:resource="#Utterance"/>
106       <rdfs:domain rdf:resource="#Move"/>
107       <rdf:type rdf:resource="http://www.w3.org/2002/07/owl#
              FunctionalProperty"/>
108     </owl:ObjectProperty>
109     <owl:ObjectProperty rdf:ID="requires">
110       <rdfs:range>
111         <owl:Class>
112           <owl:unionOf rdf:parseType="Collection">
113             <owl:Class rdf:about="#Semantic"/>
114             <owl:Class rdf:about="#Belief"/>
115           </owl:unionOf>
116         </owl:Class>
117       </rdfs:range>
118       <rdfs:domain rdf:resource="#Agenda"/>
119     </owl:ObjectProperty>
120     <owl:ObjectProperty rdf:ID="mustnot">
121       <rdfs:range>
122         <owl:Class>
123           <owl:unionOf rdf:parseType="Collection">
124             <owl:Class rdf:about="#Semantic"/>
125             <owl:Class rdf:about="#Belief"/>
126           </owl:unionOf>
127         </owl:Class>
128       </rdfs:range>
129       <rdfs:domain rdf:resource="#Agenda"/>
130     </owl:ObjectProperty>
131     <owl:ObjectProperty rdf:ID="contrarySemantic">
132       <rdfs:domain rdf:resource="#Move"/>
133       <rdfs:range rdf:resource="#Semantic"/>
134     </owl:ObjectProperty>
135     <owl:ObjectProperty rdf:ID="next">
136       <rdfs:range rdf:resource="#Agenda"/>
137       <rdfs:domain rdf:resource="#Agenda"/>
138     </owl:ObjectProperty>
139     <owl:ObjectProperty rdf:ID="hasBelief">
140       <rdfs:domain rdf:resource="#BeliefSpace"/>
141       <rdfs:range rdf:resource="#Belief"/>
142     </owl:ObjectProperty>
143     <owl:ObjectProperty rdf:ID="grammar">
144       <rdfs:range rdf:resource="#Grammar"/>
145       <rdf:type rdf:resource="http://www.w3.org/2002/07/owl#
              FunctionalProperty"/>
146       <rdfs:domain rdf:resource="#Move"/>
147     </owl:ObjectProperty>
148     <owl:DatatypeProperty rdf:ID="utteranceString">
149       <rdfs:domain rdf:resource="#Utterance"/>
150       <rdfs:range rdf:resource="http://www.w3.org/2001/XMLSchema#
              string"/>
151     </owl:DatatypeProperty>
152     <owl:DatatypeProperty rdf:ID="grammarString">
153       <rdfs:domain rdf:resource="#Grammar"/>
154       <rdfs:range rdf:resource="http://www.w3.org/2001/XMLSchema#
              string"/>
155     </owl:DatatypeProperty>
156     <owl:DatatypeProperty rdf:ID="semanticString">
157       <rdfs:domain rdf:resource="#Semantic"/>
158       <rdfs:range rdf:resource="http://www.w3.org/2001/XMLSchema#
              string"/>
159     </owl:DatatypeProperty>
160     <owl:DatatypeProperty rdf:ID="priority">
161       <rdf:type rdf:resource="http://www.w3.org/2002/07/owl#
              FunctionalProperty"/>
162       <rdfs:domain>
163         <owl:Class>
```

```
164          <owl:unionOf rdf:parseType="Collection">
165            <owl:Class rdf:about="#Agenda"/>
166            <owl:Class rdf:about="#Move"/>
167          </owl:unionOf>
168        </owl:Class>
169      </rdfs:domain>
170      <rdfs:range rdf:resource="http://www.w3.org/2001/XMLSchema#int
            "/>
171    </owl:DatatypeProperty>
172    <owl:FunctionalProperty rdf:ID="defaultValue">
173      <rdfs:range rdf:resource="http://www.w3.org/2001/XMLSchema#
            string"/>
174      <rdf:type rdf:resource="http://www.w3.org/2002/07/owl#
            DatatypeProperty"/>
175      <rdfs:domain rdf:resource="#Variable"/>
176    </owl:FunctionalProperty>
177    <owl:FunctionalProperty rdf:ID="variabledefault">
178      <rdfs:domain rdf:resource="#Belief"/>
179      <rdfs:range rdf:resource="#Variable"/>
180      <rdf:type rdf:resource="http://www.w3.org/2002/07/owl#
            ObjectProperty"/>
181    </owl:FunctionalProperty>
182    <owl:FunctionalProperty rdf:ID="variableOperator">
183      <rdfs:domain>
184        <owl:Class>
185          <owl:unionOf rdf:parseType="Collection">
186            <owl:Class rdf:about="#Agenda"/>
187            <owl:Class rdf:about="#Move"/>
188          </owl:unionOf>
189        </owl:Class>
190      </rdfs:domain>
191      <rdf:type rdf:resource="http://www.w3.org/2002/07/owl#
            DatatypeProperty"/>
192      <rdfs:range rdf:resource="http://www.w3.org/2001/XMLSchema#
            string"/>
193    </owl:FunctionalProperty>
194    <owl:FunctionalProperty rdf:ID="variableValue">
195      <rdf:type rdf:resource="http://www.w3.org/2002/07/owl#
            DatatypeProperty"/>
196      <rdfs:range rdf:resource="http://www.w3.org/2001/XMLSchema#
            string"/>
197      <rdfs:domain rdf:resource="#Belief"/>
198    </owl:FunctionalProperty>
199    <owl:FunctionalProperty rdf:ID="inWorkspace">
200      <rdfs:range rdf:resource="#WorkSpace"/>
201      <rdfs:domain rdf:resource="#History"/>
202      <rdf:type rdf:resource="http://www.w3.org/2002/07/owl#
            ObjectProperty"/>
203    </owl:FunctionalProperty>
204    <owl:FunctionalProperty rdf:ID="forAgenda">
205      <rdfs:domain rdf:resource="#History"/>
206      <rdf:type rdf:resource="http://www.w3.org/2002/07/owl#
            ObjectProperty"/>
207      <rdfs:range rdf:resource="#Agenda"/>
208    </owl:FunctionalProperty>
209    <owl:FunctionalProperty rdf:ID="agendaPriority">
210      <rdf:type rdf:resource="http://www.w3.org/2002/07/owl#
            DatatypeProperty"/>
211      <rdfs:domain rdf:resource="#History"/>
212      <rdfs:range rdf:resource="http://www.w3.org/2001/XMLSchema#int
            "/>
213    </owl:FunctionalProperty>
214    <owl:FunctionalProperty rdf:ID="addTime">
215      <rdf:type rdf:resource="http://www.w3.org/2002/07/owl#
            DatatypeProperty"/>
216      <rdfs:range rdf:resource="http://www.w3.org/2001/XMLSchema#int
            "/>
```

```
217        <rdfs:domain rdf:resource="#History"/>
218      </owl:FunctionalProperty>
219      <owl:FunctionalProperty rdf:ID="isMasterBool">
220        <rdf:type rdf:resource="http://www.w3.org/2002/07/owl#
               DatatypeProperty"/>
221        <rdfs:domain rdf:resource="#Agenda"/>
222        <rdfs:range rdf:resource="http://www.w3.org/2001/XMLSchema#
               boolean"/>
223      </owl:FunctionalProperty>
224      <owl:FunctionalProperty rdf:ID="procTime">
225        <rdfs:domain rdf:resource="#History"/>
226        <rdf:type rdf:resource="http://www.w3.org/2002/07/owl#
               DatatypeProperty"/>
227        <rdfs:range rdf:resource="http://www.w3.org/2001/XMLSchema#int
               "/>
228      </owl:FunctionalProperty>
229      <owl:FunctionalProperty rdf:ID="respawn">
230        <rdfs:range rdf:resource="http://www.w3.org/2001/XMLSchema#
               boolean"/>
231        <rdf:type rdf:resource="http://www.w3.org/2002/07/owl#
               DatatypeProperty"/>
232        <rdfs:domain rdf:resource="#Agenda"/>
233      </owl:FunctionalProperty>
234      <owl:AnnotationProperty rdf:ID="domainName">
235        <rdfs:range rdf:resource="http://www.w3.org/2001/XMLSchema#
               string"/>
236        <rdf:type rdf:resource="http://www.w3.org/2002/07/owl#
               DatatypeProperty"/>
237        <rdfs:domain rdf:resource="#DialogueDomain"/>
238      </owl:AnnotationProperty>
239    </rdf:RDF>
```

Listing B.1 The Spoken Dialogue Ontology described using OWL

Appendix C
A Scripting Language as OWL Extension

For more complex relations, such as conditional clauses, OWL is not fully appropriate: in OWL each logical operation requires its own specific relation, which would make the dialogue model needlessly complex. The fundamental logic that underlies a dialogue is defined using OWL relations that are presented in the previous section. However, in order to allow the dialogue designer to define more specific operators such as counters, conditions, or other data-manipulating commands we have implemented a scripting language that allows to define logical operations on variables. These operations are carried out if a specific move or an agenda is being processed. Three fundamental types of operations are covered by the scripting language:

- "IF/THEN" may be used to define a conditional command.
- "SET" may be used to set the value of a variable within an SDO that is currently within the focus of the ASDM. Variables of other dialogue domains might also be addressed by using the unique identifier of the SDO as prefix.
- "REQUIRES" may be used to define more complex, e.g. conditional *require* relations that exceed the expressiveness of the OWL-based relation of the Agenda class.

If necessary, the logical operators may be combined as well. The approach allows, for example, to define conditional requirements such as "agenda A REQUIRES semantic S1 IF userName == Suki". This expressions prescribes that agenda A only needs sematic S1 to be shared between user and system in case the name of the user equals "Suki". Table C.1 lists all logical operators the OwlSpeak ASDM supports. In the following paragraphs we present more examples in order to illustrate the capabilities of the scripting language.

The scripting approach allows the access of variables that are part of external SDOs. This access could also be defined via OWL—but if a relation to an external Ontology B is described within Ontology A, Ontology A cannot be parsed correctly without accessing Ontology B. This obligation defined by the OWL standard is in contrast to our approach towards a dynamic composition of various SDOs during

T. Heinroth and W. Minker, *Introducing Spoken Dialogue Systems into Intelligent Environments*, DOI 10.1007/978-1-4614-5383-3,

Table C.1 The logical operators that are supported by the scripting language

Operator name	Symbol	Logical operation
Equality	$==$	if$(a == b)$ return true
Inequality	$!=$	if$(a! = b)$ return true
And	$\&\&$	if$(a\&b)$ return true
Or	$\vert\vert$	if$(a\vert b)$ return true
Addition	$++$	return$(a+b)$
Subtraction	$--$	return$(a-b)$
Multiplication	$**$	return$(a*b)$
Division	$//$	return(a/b)
Less-than	$<<$	if$(a < b)$ return true
Greater-than	$>>$	if$(a > b)$ return true
Less-than-or-equal	$<=$	if$(a <= b)$ return true
Greater-than-or-equal	$>=$	if$(a >= b)$ return true

runtime. In fact we want to loosely couple two or more ontologies that may also be parsed even if only one of the ontologies is accessible by the ASDM. The application scenario presented in Sect. 3.5 consists of a main task: the user wants to prepare a dinner. Two ontologies may be designed for this purpose: one for discussing the menu and one for the cooking task itself. If both ontologies are within the scope of the ASDM, specific data, for example, the required ingredients have to be imported into the cooking ontology in order to prevent the user from entering the same input twice. The variables of the imported ontology may be addressed by using the unique domain name as prefix. The uniqueness of the domain names (i.e. the ontology names) is a hard requirement that allows the ASDM to distinguish between the different dialogue descriptions in an unambiguous way. However, if only one of those ontologies is currently activated, i.e. if the user wants to cook without discussing a menu, it may also be interpreted by the ASDM in a "stand-alone" manner. Besides the fundamental operations described above, several basic logical operators are supported by the scripting language as well (see Table C.1). The following examples provide a more detailed idea of the approach:

- If a specific agenda should only be processed if a variable has been set to a specific value, the variableOperator:String field may provide an operation such as

  $$REQUIRES(\%dialog\_progress\% == phase1).$$

  In this example "dialogue_progress" is the name of the variable and "phase1" the value it should have. Any constellations and numbers of REQUIRES operations may be cumulated using the logical operators listed in Table C.1.
- If a variable should be set to a specific value, the variableOperator:String field may provide an operation such as

  $$SET(main\_course = fish);(dialog\_progress = phase2).$$

In this example "main_course" and "dialogue_progress" are the variables that should be set to the corresponding value in case the regarding move has been processed. Any numbers of SET operations may be added using a semicolon as separator.

- If a variable operation should only be processed depending on a logical operation evaluated to true, the "IF/THEN" operator may be utilised. A valid expression would be, for example,

$$IF(\%dialog\_progress\% == phase2)THEN(main\_course = fish).$$

Indicated by two "%" the first variable will be replaced with its current value. If the value is equal to the value of the second variable ("phase2") the value of the variable "main_course" will be set to "fish".

All logical operators in the examples above may be replaced by those listed in Table C.1 making the scripting language a powerful extension to the relations encoded within the OWL-based SDO.

Appendix D
The Major Aspects of the Dialogue Decision Logic

In this chapter we provide two UML Activity diagrams showing the most important decision processes of the OwlSpeak ASDM. Figure D.1 shows the process of deciding which Agenda the ASDM has to select to be used for the upcoming dialogue generation.

This decision process is one part of the dialogue logic implemented as part of the presenter layer. Before the actual agenda is detected, the system also decides which SDO has to be used for the dialogue generation. After the correct ontology and the correct agenda are being selected, the ASDM has to decide which moves (utterances and/or grammars) have to be used to finally render the view that reflects the dialogue. In case the system has to process an input on part of the user it is necessary to integrate the newly gathered information into the SDO and to update the user workspace that contains all agendas that still have to be processed. Figure D.2 shows this process as UML Activity diagram.

T. Heinroth and W. Minker, *Introducing Spoken Dialogue Systems into Intelligent Environments*, DOI 10.1007/978-1-4614-5383-3,

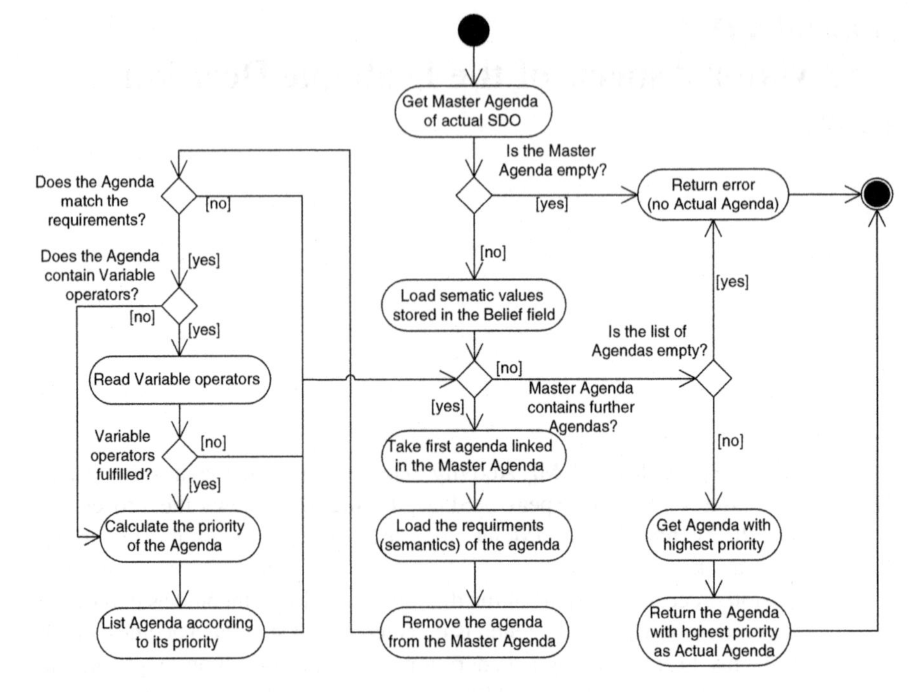

Fig. D.1 UML Activity Diagram showing the "GetActualAgenda" decision process

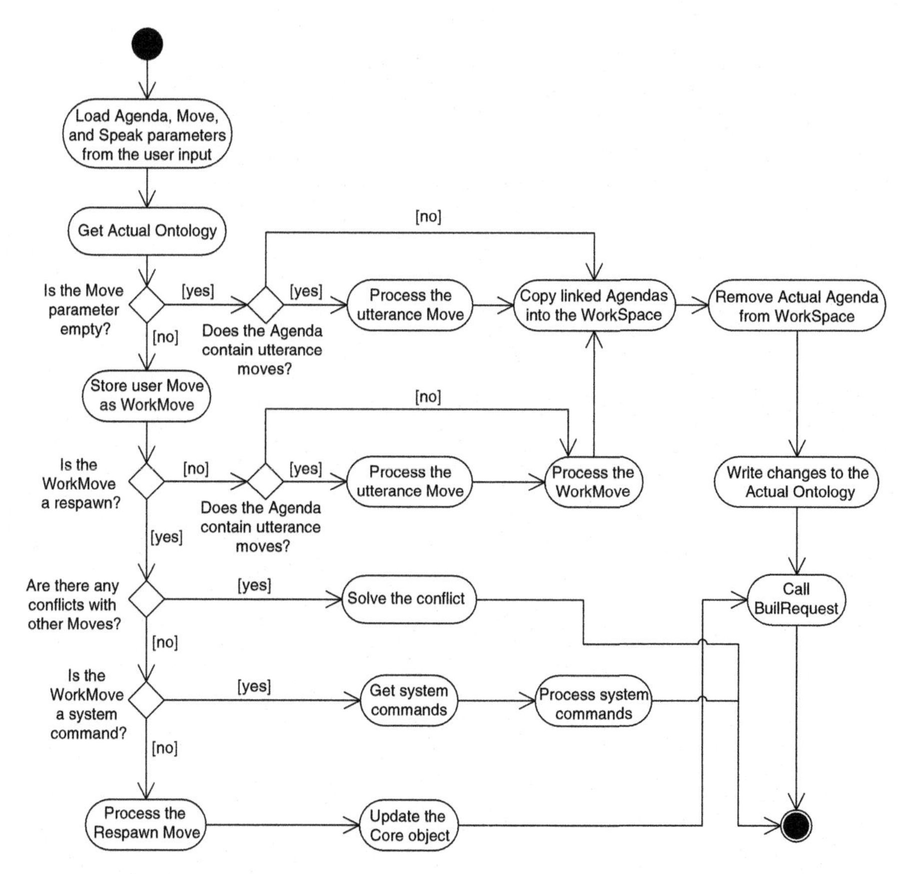

Fig. D.2 UML Activity Diagram showing the "ProcessAgenda" functionality that integrates the information of a user input into the SDO

Appendix E
Questionnaire

An exemplary questionnaire deployed for the evaluation sessions is shown in Figs. E.1–E.3. An unmodified version has been used for the repair strategies experiment that is presented in Sect. 5.4. The other evaluation session were conducted with similar versions that were tailored to the specific target of investigation. Figure E.1 consists of a questionnaire ID, the date, the time, and personal information about the subject. The second part is shown in Fig. E.2 and focusses on the users' rating of the spoken dialogue itself. This rating is somehow difficult since it is not clearly evident for the subject where the SDS ends and at which point the ASDM begins. The third part is shown in Fig. E.3. This part consists of SASSI-based statements (Hone and Graham 2000). The subjects can agree or disagree with the statements within a specific range.

Here, we focussed on statements related to dialogue management since this was the main interest of our investigations. Modified versions of this questionnaire have also been used for the initial system evaluation (Sect. 5.1), the evaluation of the topic switching strategies (Sect. 5.3), and the evaluation of the advanced understanding methods (Sect. 5.6).

T. Heinroth and W. Minker, *Introducing Spoken Dialogue Systems into Intelligent Environments*, DOI 10.1007/978-1-4614-5383-3,

Fragebogen – Sprachdialog

Testperson Nr.: ... | Datum: ...... | Uhrzeit:......

1. Persönliche Angaben:

	männlich	weiblich			
Geschlecht:	□	□			
	≤25	26-35	36-45	46-55	≥56
Alter:	□	□	□	□	□
	Anfänger				Experte
Umgang mit Computern:	□	□	□	□	□

Fig. E.1 First part of the exemplary SASSI-based questionnaire as it has been used for the evaluation series

2. Allgemeine Fragen:

	Ja	Nein			
Konnten sie den Dialog erfolgreich beenden?	□	□			
	Gut				Schlecht
Wie war das System zu verstehen?	□	□	□	□	□
	Nie				Immer
Wurden Sie vom System verstanden?	□	□	□	□	□
Funktionierte das System wie erwartet?	□	□	□	□	□
Wussten Sie immer was Sie sagen können?	□	□	□	□	□

Fig. E.2 Second part of the exemplary SASSI-based questionnaire as it has been used for the evaluation series

3. Zutreffende Aussagen:

	Gar nicht	Eher nicht	Teils teils	Eher zu	Voll zu
Das System ist nicht zuverlässig.	□	□	□	□	□
Das System ist benutzerfreundlich.	□	□	□	□	□
Das System macht wenig Fehler.	□	□	□	□	□
Ich konnte Fehler einfach korrigieren.	□	□	□	□	□
Die Interaktion mit dem System ist unvorhersehbar.	□	□	□	□	□
Der Umgang mit dem System ist angenehm.	□	□	□	□	□
Die Interaktion mit dem System ist irritierend.	□	□	□	□	□
Ich war nicht immer sicher, was das System tat.	□	□	□	□	□
Die Interaktion mit dem System läuft schnell.	□	□	□	□	□
Ich musste mich stark konzentrieren.	□	□	□	□	□
Die Interaktion mit dem System ist effektiv.	□	□	□	□	□

Fig. E.3 Third part of the exemplary SASSI-based questionnaire as it has been used for the evaluation series

References

Abowd, G., Atkeson, C., & Essa, I. (1998). Ubiquitous smart spaces. Technical report, DARPA.

Axelsson, J., Cross, C., Lie, H. W., McCobb, G., Raman, T. V., & Wilson, L. (2001). Xhtml+voice profile 1.0. Technical report, W3C.

Bachmann, P. (1894). *Die analytische Zahlentheorie*, vol. 2. Leipzig: Teubner.

Baum, L. E., Petrie, T., Soules, G., & Weiss, N. (1970). A maximization technique occurring in the statistical analysis of probabilistic functions of markov chains. *The Annals of Mathematical Statistics*, *41*(1), 164–171.

Bechhofer, S., Volz, R., & Lord, P. (2003). Cooking the semantic web with the owl api. In *The Semantic Web – ISWC 2003*, (pp. 659–675). Springer.

Bellik, Y., Pruvost, G., Martin, J.-C., Tan, N., Minker, W., & Heinroth, T. (2010). D16 – user interaction adaptation component. Confidential deliverable, The ATRACO Project (FP7/2007–2013 grant agreement no:216837).

Berton, A., Bühler, D., & Minker, W. (2006). *SmartKom-Mobile Car: User Interaction with Mobile Services in a Car Environment* (SmartKom: Foundations of Multi-Modal Dialogue Systems ed.)., (pp. 523–541). Cognitive Technologies. Heidelberg: Springer.

Beslay, L., & Hakala, H. (2007). Digital territory: Bubbles. In P. T. Kidd (Ed.), *European visions for the knowledge age: a quest for new horizons in the information society*. Cheshire Henbury.

Bezold, M. (2011). *Adapting Multimodal Dialogue Systems to User Behaviour*. PhD thesis, Ulm University.

Bidot, J., Goumopoulos, C., & Calemis, I. (2011). Using ai planning and late binding for managing service workflows in intelligent environments. In *Proc. of the International Conference on Pervasive Computing and Communications (PerCom)*, (pp. 156–163). IEEE.

Black, A. W., Burger, S., Conkie, A., Hastie, H. W., Keizer, S., Lemon, O., Merigaud, N., Parent, G., Schubiner, G., Thomson, B., Williams, J. D., Yu, K., Young, S., & Eskenazi, M. (2011). Spoken dialog challenge 2010: Comparison of live and control test results. In *SIGDIAL Conference*, (pp. 2–7).

Bohlin, P., Bos, J., Larsson, S., Lewin, I., Matheson, C., & Milward, D. (1999). Survey of existing interactive systems – trindi deliverable d1.3. Technical report, Gothenburg University.

Bohus, D., Raux, A., Harris, T. K., Eskenazi, M., & Rudnicky, E. I. (2007). Olympus: an open-source framework for conversational spoken language interface research. In *HLT-NAACL 2007 workshop on Bridging the Gap: Academic and Industrial Research in Dialog Technology*.

Bohus, D., & Rudnicky, A. (2002). Integrating multiple knowledge sources for utterance-level confidence annotation in the cmu communicator spoken dialog system. Technical report, Roots in the Town. In 2nd International Workshop on Community Networking. 1995. Princeton, NJ: IEEE Communications Society

Bohus, D., & Rudnicky, A. (2005). Sorry i didn't catch that: An investigation of non-understanding errors and recovery strategies. In *Proceedings of SIGdial-2005*, Lisbon, Portugal.

T. Heinroth and W. Minker, *Introducing Spoken Dialogue Systems into Intelligent Environments*, DOI 10.1007/978-1-4614-5383-3,

Bohus, D., & Rudnicky, A. I. (2009). The ravenclaw dialog management framework: Architecture and systems. *Computer Speech & Language*, *23*, 332–361.

Brown, M., Burnett, D., Candell, E., Carter, J., Dahl, D., Ghosh, D., Hunt, A., Krause, S., Lerner, S., Lucas, B., Marschner, J., McGlashan, S., Normandin, Y., Porter, B., Raggett, D., Ramsthaler, D., Tichelen, L. V., Wang, K., & Werner, L. (2004). Speech recognition grammar specification version 1.0. Technical report, W3C.

Bühler, D. (2009). *Towards Domain-driven Dialogue - Application Control and Problem Solving*. PhD thesis, Ulm University.

Burkhardt, F., Huber, R., & Batliner, A. (2007). Application of speaker classification in human machine dialog systems. In *Speaker Classification I: Fundamentals, Features, and Methods*, (pp. 174–179). Berlin, Heidelberg: Springer.

Burkhardt, F., Metze, F., & Stegmann, J. (2008). *Speaker classification for next-generation voice-dialog systems*, (pp. 497–528). Wiley.

Cáceres, M. (2011). Widget packaging and configuration (working draft). Technical report, W3C.

Chin, J., Diehl, V., & Norman, K. (1988). Development of an instrument measuring user satisfaction of the human–computer interface. In *Proceedings of ACM CHI 88 Conference on Human Factors in Computing*, (pp. 213–218).

Chomsky, N. (1956). Three models for the description of language. *IRE Transactions on Information Theory*, *2*, 113–124.

Chung, G., Seneff, S., Wang, C., & Hetherington, L. (2004). A dynamic vocabulary spoken dialogue interface. In *Proc. ICSLP*, (pp. 1457–1460).

Clark, H. H., & Schaefer, E. F. (1989). Contributing to discourse. *Cognitive Science*, *13*(2), 259–294.

Colmerauer, A., & Roussel, P. (1996). The birth of prolog. In T. J. Bergin, Jr., & R. G. Gibson, Jr. (Eds.), *History of programming languages—II* (pp. 331–367). New York, NY, USA: ACM.

Cook, D., Youngblood, M., & Das, S. (2006). A multi-agent approach to controlling a smart environment. In J. Augusto and C. Nugent (Eds.), *Designing Smart Homes*, vol. 4008 of *Lecture Notes in Computer Science* (pp. 165–182). Heidelberg: Springer.

Cornelius, R. (1996). *The science of emotion : research and tradition in the psychology of emotions*. Upper Saddle River, NJ, USA: Prentice Hall.

Coutaz, J., Crowley, J., Dobson, S., & Garlan, D. (2005). Context is key. *Communications of the ACM*, *48*(3), 49–53.

Cowie, R., Douglas-Cowie, E., Tsapatsoulis, N., Votsis, G., Kollias, S., Fellenz, W., & Taylor, J. G. (2001). Emotion recognition in human–computer interaction. *Signal Processing Magazine*, *18*(1), 32–80.

Daniels, J. (2000). Integrating a spoken language system with agents for operational information access. In *AAAI*, (pp. 1002–1007).

Dervin, B., Foreman-Wernet, L., & Lauterbach, E. (2003). *Sense-making methodology reader: Selected writings of Brenda Dervin*. Hampton Press Inc.

Dretske, F. (1991). *Explaining behavior: Reasons in a world of causes*. Cambridge, MA, USA: MIT.

Duong, T., Bui, H., Phung, D., & Venkatesh, S. (2005). Activity recognition and abnormality detection with the switching hidden semi-markov model. In *Computer Vision and Pattern Recognition, 2005. CVPR 2005. IEEE Computer Society Conference on*, vol. 1, (pp. 838–845). IEEE.

Fahrmeir, L., Hamerle, A., & Tutz, G. (1984). *Multivariate statistische Verfahren*. New York: Walter de Gruyter.

Ferguson, G., Allen, J., Blaylock, N., Byron, D., Chambers, N., Dzikovska, M., Galescu, L., Shen, X., Swier, R., & Swift, M. (2002). The Medication Advisor Project: Preliminary report. Technical Report TR776, University of Rochester Computer Science Department.

Fowler, M. (2006). Passive view.

Franke, J., Daniels, J., & McFarlane, D. (2002). Recovering context after interruption. In *CogSci'02*, (pp. 310–315).

Garrett, J. J. (2005). Ajax: A new approach to web applications. http://adaptivepath.com/ideas/essays/archives/000385.php.
Gervasio, M., & Murdock, J. (2009). What were you thinking?: filling in missing dataflow through inference in learning from demonstration. In *Proceedings of the 14th international conference on Intelligent user interfaces*, (pp. 157–166). ACM.
Gil, Y., & Ratnakar, V. (2008). Towards intelligent assistance for to-do lists. In *Proceedings of the 13th international conference on Intelligent user interfaces*, (pp. 329–332). ACM.
Ginzburg, J., & Cooper, R. (2004). Clarification, ellipsis, and the nature of contextual updates in dialogue. *Linguistics and Philosophy*, *27*(3), 297–365.
Gnjatović, M., & Rösner, D. (2008). Adaptive dialogue management in the nimitek prototype system. In *Proceedings of the 4th IEEE PIT workshop*, (pp. 14–25). Berlin, Heidelberg: Springer.
Goumopoulos, C., & Kameas, A. (2009). Ambient ecologies in smart homes. *The Computer Journal*, *52*(8), 922–937.
Habibi, M., Rahbar, S., & Sameti, H. (2010). Divided pomdp method for complex menu problems in spoken dialogue systems. In *Spoken Language Technology Workshop (SLT), 2010 IEEE*, (pp. 484–489). IEEE.
Hamp, B., & Feldweg, H. (1997). Germanet – a lexical-semantic net for german. In *Proceedings of ACL workshop Automatic Information Extraction and Building of Lexical Semantic Resources for NLP Applications*, (pp. 9–15). Citeseer.
Heinroth, T., & Denich, D. (2011). Spoken Interaction within the Computed World: Evaluation of a Multitasking Adaptive Spoken Dialogue System. In *35th Annual IEEE International Computer Software and Applications Conference (COMPSAC 2011)*. IEEE.
Heinroth, T., Denich, D., & Schmitt, A. (2010). Owlspeak - adaptive spoken dialogue within intelligent environments. In *8th IEEE International Conference on Pervasive Computing and Communications Workshops (PERCOM Workshops)*, (pp. 666 – 671). Mannheim, Germany.
Heinroth, T., Grotz, M., Nothdurft, F., & Minker, W. (2012). Adaptive speech recognition for intuitive model-based spoken dialogues. In *Proceedings of the Eighth Conference on International Language Resources and Evaluation (LREC'12)*. European Language Resources Association (ELRA).
Heinroth, T., Koleva, S., & Minker, W. (2011). Topic switching strategies for spoken dialogue systems. In *Proc. of the 12th Annual Conference of the International Speech Communication Association*.
Heinroth, T., & Minker, W. (Eds.). (2011). *Next Generation Intelligent Environments: Ambient Adaptive Systems*. Boston, USA: Springer.
Herm, O., Schmitt, A., & Liscombe, J. (2008). When calls go wrong: How to detect problematic calls based on log-files and emotions? In *Proc. of the International Conference on Speech and Language Processing (ICSLP)*.
Hildebrand, A., & Sá, V. (2000). Embassi: electronic multimedia and service assistance. In *oceedings of the Internet Measurement Conference (IMC)*, (pp. 50–59).
Hone, K. S., & Graham, R. (2000). Towards a tool for the subjective assessment of speech system interfaces (sassi). *Natural Language Engineering*, *6*, 287–305.
Horridge, M., Bechhofer, S., & Noppens, O. (2007). Igniting the owl 1.1 touch paper: The owl api. In *Proc. OWL-ED*, vol. 258.
Huerta, J. M. (2000). *Robust Speech Recognition in GSM Mobile Environments*. PhD thesis, Carnegie Mellon University.
Hunt, A. (2000). Jspeech grammar format. W3C Note http://www.w3.org/TR/jsgf/.
Intille, S. S., Larson, K., Beaudin, J. S., Tapia, M., Kaushik, P., Nawyn, J., and Mcleish, T. J. (2005). The placelab: a live-in laboratory for pervasive computing research (video. In *Proceedings of Pervasive 2005 Video Program*.
ISO (2008). Iso/iec 29341–2:2008 information technology – upnp device architecture – part 2: Basic device control protocol - basic device. Technical report, INTERNATIONAL ORGANIZATION FOR STANDARDIZATION.

ITU (2005). Parameters describing the interaction with spoken dialogue systems. ITU-T Recommendation Supplement 24 to P-Series, International Telecommunication Union, Geneva, Switzerland. Based on ITU-T Contr. COM 12–17 (2009).
Jiang, H. (2005). Confidence measures for speech recognition: A survey. *Speech Communication, 45*(4), 455–470.
Johnston, M., Baggia, P., Burnett, D., Carter, J., Dahl, D., & McCobb, G. (2009). Emma: Extensible multimodal annotation markup language; World Wide Web Consortium Recommendation REC-emma-2009021. Technical report, W3C.
Jokinen, K., Kerminen, A., Kaipainen, M., Jauhiainen, T., Wilcock, G., Turunen, M., Hakulinen, J., Kuusisto, J., & Lagus, K. (2002). Adaptive dialogue systems-interaction with interact. In *Proceedings of the 3rd SIGdial workshop on Discourse and dialogue-Volume 2*, (pp. 64–73). ACL.
Jurafsky, D., & Martin, J. H. (2000). *Speech and Language Processing: An Introduction to Natural Language Processing, Computational Linguistics and Speech Recognition (Prentice Hall Series in Artificial Intelligence)* (1st ed.). Prentice Hall.
Kaelbling, L., Littman, M., & Cassandra, A. (1998). Planning and acting in partially observable stochastic domains. *Artificial Intelligence, 101*(1–2), 99–134.
Kientz, J. A., Patel, S. N., Jones, B., Price, E., Mynatt, E. D., & Abowd, G. D. (2008). The georgia tech aware home. In *CHI '08 extended abstracts on Human factors in computing systems*, CHI EA '08, (pp. 3675–3680). New York, NY, USA: ACM.
Kleene, S. (1988). *Introduction to metamathematics*. Wolters-Noordhoff.
Knuth, D. E. (1964). Backus normal form vs. Backus Naur form. *Communications of the ACM, 7*(12), 735–736.
Konings, B., & Schaub, F. (2011). Territorial privacy in ubiquitous computing. In *Wireless On-Demand Network Systems and Services (WONS), 2011 Eighth International Conference on*, (pp. 104–108). IEEE.
Könings, B., Wiedersheim, B., & Weber, M. (2011). Privacy & trust in ambient intelligence environments. In W. Minker and T. Heinroth (Eds.), *Next Generation Intelligent Environments* (pp. 227–252). New York: Springer.
Krasner, G., & Pope, S. (1998). A cookbook for using the model-view-controller user interface paradigm in smalltalk-80. *Journal of Object-Oriented Programming, 1*(3), 26–49.
Kruskal, W., & Wallis, W. (1952). Use of ranks in one-criterion variance analysis. *Journal of the American statistical Association, 47*(260), 583–621.
Larsson, S. (2002). *Issue-based Dialogue Management*. PhD thesis, Göteborg University, Sweden.
Larsson, S., & Traum, D. (2000). Information state and dialogue management in the trindi dialogue move engine. *Natural Language Engineering Special Issue, 6*, 323–340.
Levenshtein, V. (1966). Binary codes capable of correcting deletions, insertions, and reversals. *Soviet Physics Doklady, 10*(8), 707–710.
Lewis, J. R. (1995). Ibm computer usability satisfaction questionnaires: Psychometric evaluation and instructions for use. *International Journal of Human–Computer Interaction, 7*(1), 57–78.
Limbourg, Q., Vanderdonckt, J., Michotte, B., Bouillon, L., & López-Jaquero, V. (2005). Usixml: A language supporting multi-path development of user interfaces. In *9th IFIP Working Conference on Engineering for Human–Computer Interaction*, (pp. 134–135). Springer.
Litman, D., & Pan, S. (2002). Designing and evaluating an adaptive spoken dialogue system. *User Modeling and User-Adapted Interaction, 12*(2), 111–137.
Lockwood, S., & Cook, D. (2008). Computer, light on! In *The 4th IET International Conference on Intelligent Environments*, Seattle, USA.
López-Cózar, R., & Callejas, Z. (2006). Two-level speech recognition to enhance the performance of spoken dialogue systems. *Knowledge-Based Systems, 19*(3), 153–163.
López-Cózar, R., & Callejas, Z. (2008). Asr post-correction for spoken dialogue systems based on semantic, syntactic, lexical and contextual information. *Speech Communication, 50*(8–9), 745–766.
López-Cózar, R., & Callejas, Z. (2010). *Multimodal dialogue for ambient intelligence and smart environments*, chapter 21, (pp. 559–579). Springer.

Mankiewicz, R. (2000). *The story of mathematics*. Princeton: Princeton University Press.
Mann, H., & Whitney, D. (1947). On a test of whether one of two random variables is stochastically larger than the other. *The annals of mathematical statistics*, *18*(1), 50–60.
McFarlane, D. (2002). Comparison of four primary methods for coordinating the interruption of people in human–computer interaction. *Human–Computer Interaction*, *17*, 63–139.
McGuinness, D. L., & van Harmelen, F. (2004). Owl web ontology language. Technical report, W3C.
McTear, M. (2004). *Spoken Dialogue Technology: Toward the Conversational User Interface*. London: Springer.
McTear, M., O'Neill, I., Hanna, P., Liu, X., McTear, M., O'Neill, I., Hanna, P., & Liu, X. (2005). Handling errors and determining confirmation strategies–an object-based approach. *Speech Communication*, *45*(3), 249–269. Special Issue on Error Handling in Spoken Dialogue Systems.
Metze, F., Englert, R., Bub, U., Burkhardt, F., & Stegmann, J. (2008). Getting closer: tailored human–computer speech dialog. *Universal Access in the Information Society*, *8*, 97–108.
Miller, G. (1956). The magical number seven, plus or minus two: some limits on our capacity for processing information. *Psychological review*, *63*(2), 81–97.
Miller, G. (1995). Wordnet: a lexical database for english. *Communications of the ACM*, *38*(11), 39–41.
Minker, W., López-Cózar, R., & McTear, M. (2009). The role of spoken language dialogue interaction in intelligent environments. *Journal of Ambient Intelligence and Smart Environments*, *1*(1), 31–36.
Montoro, G., Alamán, X., & Haya, P. A. (2004). Spoken interaction in intelligent environments: A working system. In *Advances in Pervasive Computing*.
Mozer, M. C. (2005). *Lessons from an Adaptive Home*, (pp. 271–294). Wiley.
Nakano, M., Miyazaki, N., Hirasawa, J.-i., Dohsaka, K., & Kawabata, T. (1999). Understanding unsegmented user utterances in real-time spoken dialogue systems. In *Proceedings of the 37th annual meeting of the Association for Computational Linguistics on Computational Linguistics*, ACL '99, (pp. 200–207). Stroudsburg, PA, USA: ACL.
Nevin, B., & Johnson, S. (2002). *The legacy of Zellig Harris: language and information into the 21st century*. John Benjamins Publishing Company.
Niezen, G., van der Vlist, B., Hu, J., & Feijs, L. (2010). From events to goals: Supporting semantic interaction in smart environments. In *2010 IEEE Symposium on Computers and Communications (ISCC)*, (pp. 1029–1034). IEEE.
Nuance (2008). Nuance speech recognition system version 8.5 grammar developer's guide. Technical report, Nuance Communications. visited 05.09.2010.
Oh, A. H., & Rudnicky, A. I. (2000). Stochastic language generation for spoken dialogue systems. In *Proceedings of the 2000 ANLP/NAACL Workshop on Conversational systems - Volume 3*, ANLP/NAACL-ConvSyst '00, (pp. 27–32). Stroudsburg, PA, USA: ACL.
Oshry, M., Auburn, R., Baggia, P., Bodell, M., Burke, D., Burnett, D. C., Candell, E., Carter, J., McGlashan, S., Lee, A., Porter, B., & Rehor, K. (2007). Voice extensible markup language (voicexml) 2.1. Technical report, W3C.
Paternò, F., Mancini, C., & Meniconi, S. (1997). Concurtasktrees: A diagrammatic notation for specifying task models. In *Proceedings of the IFIP TC13 Interantional Conference on Human–Computer Interaction*, (pp. 362–369).
Pittermann, J. (2008). *Speech-Emotion Recognition in Adaptive Dialogue Systems*. PhD thesis, Ulm University.
Pittermann, J., Pittermann, A., & Minker, W. (2009). *Handling Emotions in Human–Computer Dialogues*. Dordrecht, The Netherlands: Springer.
Plutchik, R. (1980). *Emotion: A Psychoevolutionary Synthesis*. New York, USA: Harper & Row.
Potel, M. (1996). MVP: Model-View-Presenter The Taligent Programming Model for C++ and Java. Technical report, Taligent Inc.
Pruvost, G., Heinroth, T., Bellik, Y., & Minker, W. (2011). *Next Generation Intelligent Environments: Ambient Adaptive Systems*, chapter 5, (pp. 151–192). Springer.

Puerta, A., & Eisenstein, J. (2002). Ximl: a common representation for interaction data. In *Proceedings of the 7th International Conference on Intelligent User Interfaces*, (pp. 214–215). ACM.

Qu, Y. (2001). *A Constraint-Based Model of Mixed-Initiative Dialogue in Information-Seeking Interactions*. PhD thesis, School of Computer Science, Carnegie Mellon University.

Qu, Y. (2002). A constraint-based approach for cooperative information-seeking dialog. In *Proc. INLG*.

Quesada, J. F., Garcia, F., Sena, E., Bernal, J. A., & Amores, G. (2001). Dialogue management in a home machine environment: Linguistic components over an agent architecture. *Procesamiento del Lenguaje Natural*, *27*, 89–96.

Raux, A., & Eskenazi, M. (2007). A multi-layer architecture for semi-synchronous event-driven dialogue management. In *ASRU. IEEE Workshop on Automatic Speech Recognition Understanding*, (pp. 514–519).

Reenskaug, T. (1979). Models - views - controllers. Technical report, Xerox PARC.

rí Adámek, J. (2008). *Theoretische Informatik (lecture notes)*. Technische Universität Braunschweig.

Rohlicek, J., Russell, W., Roukos, S., & Gish, H. (1989). Continuous hidden Markov modeling for speaker-independent word spotting. In *ICASSP'89*, (pp. 627–630). IEEE.

Román, M., Hess, C., Cerqueira, R., Campbell, R. H., & Nahrstedt, K. (2002). Gaia: A middleware infrastructure to enable active spaces. *IEEE Pervasive Computing*, *1*, 74–83.

Ruser, H., Borodulkin, L., & Leisner, D. (2003). Multi-modal 'smart home' user interface. In *Signals Systems Decision and Information Technology (SSD)*.

Schattenberg, B., Balzer, S., & Biundo, S. (2006). Knowledge-based Middleware as an Architecture for Planning and Scheduling Systems. In *Proc. of the 16th International Conference on Automated Planning and Scheduling (ICAPS-06)*, Ambleside, The English Lake District, UK.

Schmitt, A., Heinroth, T., & Bertrand, G. (2009). Towards emotion, age- and gender-aware voicexml applications. In *5th International Conference on Intelligent Environments (IE'09)*, vol. 2 of *Ambient Intelligence and Smart Environments*, (pp. 34–41). IOS Press.

Schmitt, A., & Liscombe, J. (2008). Detecting Problematic Calls With Automated Agents. In *4th IEEE Tutorial and Research Workshop Perception and Interactive Technologies for Speech-Based Systems*, Irsee, Germany.

Schmitt, A., Schatz, B., & Minker, W. (2011). Modeling and predicting quality in spoken human–computer interaction. In *Proceedings of the SIGDIAL 2011 Conference*, (pp. 173–184). Portland, Oregon, USA: ACL.

Schnelle-Walka, D., & Feldes, S. (2009). Towards mixed-initiative concepts in smart environments. In *Proceedings of Workshop Interacting with Smart Objects*.

Seneff, S., Hurley, E., Lau, R., Pao, C., Schmid, P., & Zue, V. (1998). Galaxy-ii: A reference architecture for conversational system development. In *Proceedings of the international conference on spoken language processing*, (pp. 931–934).

Shanmugham, S., Monaco, P., & Eberman, B. (2006). A media resource control protocol (mrcp). RFC 4463 http://tools.ietf.org/html/rfc4463.

Shannon, C. (1948). A mathematical theory of communication. *Bell Systems Technical Journal*, *27*, 623–656.

Skantze, G. (2003). Exploring human error handling strategies: Implications for spoken dialogue systems. In *Proceedings of the ISCA Workshop on Error Handling in Spoken Dialogue Systems*, (pp. 71–76). Citeseer.

Sonntag, D., Engel, R., Herzog, G., Pfalzgraf, A., Pfleger, N., Romanelli, M., & Reithinger, N. (2007). *SmartWeb Handheld – Multimodal Interaction with Ontological Knowledge Bases and Semantic Web Services*, vol. 4451 of *Lecture Notes in Computer Science*, (pp. 272–295). Berlin/Heidelberg: Springer.

Stoline, M. (1981). The status of multiple comparisons: simultaneous estimation of all pairwise comparisons in one-way anova designs. *American Statistician*, *35*(3), 134–141.

Swerts, M., Litman, D., & Hirschberg, J. (2000). Corrections in spoken dialogue systems. In *Proceedings of the International Conference on Spoken Language Processing*, vol. 2, (pp. 615–618). Citeseer.

Traum, D., & Larsson, S. (2003). *The information state approach to dialogue management*, chapter 15, (pp. 325–353). Kluwer.
Turing, A. (1937). On computable numbers, with an application to the Entscheidungsproblem. *Proceedings of the London Mathematical Society*, *2*(1), 230.
Ubisense (2011). Ubisense series 7000 ip sensors. http://www.ubisense.net/en/media/pdfs/factsheets_pdf/88679_series_7000_ip_sensors_combined.pdf.
van Helvert, J., Hagras, H., & Kameas, A. (2009). D27 - prototype testing and validation (year 2). Restricted deliverable, The ATRACO Project (FP7/2007–2013 grant agreement n 216837).
van Helvert, J., Hagras, H., Wagner, C., Dooley, J., Bacon, R., & Bilgin, A. (2011). D27 - prototype testing and validation (year 3). Restricted deliverable, The ATRACO Project (FP7/2007–2013 grant agreement n 216837).
Van Welie, M., Van der Veer, G., & Eliëns, A. (1998). An ontology for task world models. In *Proceedings of DSV-IS98*, (pp. 3–5). Abingdon, UK: Springer.
Vipperla, R., Wolters, M., Georgila, K., & Renals, S. (2009). Speech input from older users in smart environments: Challenges and perspectives. In *Proceedings HCI International: Universal Access in Human–Computer Interaction. Intelligent and Ubiquitous Interaction Environments*, number 5615 in Lecture Notes in Computer Science (pp. 117–126). Springer.
Voxeo (2011). Voxeo prophecy. http://www.voxeo.com/products/.
Wagner, C., & Hagras, H. (2010). D14 – artefact operation adaptation component. Confidential deliverable, The ATRACO Project (FP7/2007–2013 grant agreement n 216837).
Walker, M., Rudnicky, A., Prasad, R., Aberdeen, J., Bratt, E., Garofolo, J., Hastie, H., Le, A., Pellom, B., Potamianos, A., et al. (2002). Darpa communicator: Cross-system results for the 2001 evaluation. In *Proc. of ICSLP*. Citeseer.
Walker, M. A., Litman, D. J., Kamm, C. A., & Abella, A. (1997). Paradise: a framework for evaluating spoken dialogue agents. In *Proceedings of the eighth conference on European chapter of the Association for Computational Linguistics*.
Wang, K. (2002). Salt: an xml application for web-based multimodal dialog management. In *Proceedings of the 2nd workshop on NLP and XML - Volume 17*, (pp. 1–8).
Ward, W., & Issar, S. (1994). Recent improvements in the cmu spoken language understanding system. In *Proceedings of the workshop on Human Language Technology, HLT '94*, (pp. 213–216). Stroudsburg, PA, USA: ACL.
Warren, W. (2006). The dynamics of perception and action. *Psychological review*, *113*(2), 358.
Wechsung, I., & Naumann, A. B. (2008). Evaluation methods for multimodal systems: A comparison of standardized usability questionnaires. *Lecture Notes in Computer Science*, *5078*, 276–284.
Williams, J., & Young, S. (2007). Scaling pomdps for spoken dialog management. *IEEE Transactions on Audio, Speech, and Language Processing*, *15*(7), 2116–2129.
Yang, F., Heeman, P., & Kun, A. (2008). Switching to real-time tasks in multi-tasking dialogue. In *COLING'08*, (pp. 1025–1032). ACL.
Yang, F., Heeman, P. A., & Kun, A. L. (2011). An investigation of interruptions and resumptions in multi-tasking dialogues. *Computational Linguistics*, *37*(1), 75–104.
Young, S. (2007). Using POMDPs for dialog management. In *Spoken Language Technology Workshop, 2006. IEEE*, (pp. 8–13). IEEE.
Young, S., Gasic, M., Keizer, S., Mairesse, F., Schatzmann, J., Thomson, B., & Yu, K. (2010). The hidden information state model: A practical framework for pomdp-based spoken dialogue management. *Computer Speech & Language*, *24*(2), 150–174.
Young, S., Williams, J., Schatzmann, J., Stuttle, M., & Weilhammer, K. (2006). D4.3: Bayes net prototype - the hidden information state dialogue manager. Technical report, TALK - Talk and Look: Tools for Ambient Linguistic Knowledge, IST-507802, 6th FP.
Zgorzelski, A., Schmitt, A., Heinroth, T., & Minker, W. (2010). Repair strategies on trial: which error recovery do users like best? In *Proc. of the International Conference on Speech and Language Processing (ICSLP)*.

Index

T. Heinroth and W. Minker, *Introducing Spoken Dialogue Systems into Intelligent Environments*, DOI 10.1007/978-1-4614-5383-3,